ENERGY SCIENCE, ENGINEERING AND TECHNOLOGY

BIODIESEL PRODUCTION TECHNOLOGIES

ENERGY SCIENCE, ENGINEERING AND TECHNOLOGY

Additional books in this series can be found on Nova's website under the Series tab.

Additional E-books in this series can be found on Nova's website under the E-books tab.

ENERGY SCIENCE, ENGINEERING AND TECHNOLOGY

BIODIESEL PRODUCTION TECHNOLOGIES

JORGE MARIO MARCHETTI

Nova Science Publishers, Inc.
New York

For permission to use material from this book please contact us:
Telephone 631-231-7269; Fax 631-231-8175
Web Site: http://www.novapublishers.com

Additional color graphics may be available in the e-book version of this book.

LIBRARY OF CONGRESS CATALOGING-IN-PUBLICATION DATA

Marchetti, Jorge Mario.
Biodiesel production technologies / author, Jorge Mario Marchetti.
p. cm.
Includes bibliographical references and index.
ISBN 978-1-61668-963-6 (hardcover)
1. Biodiesel fuels. I. Title.
TP359.B46M368 2010
662'.669--dc22
2010018336

Published by Nova Science Publishers, Inc. ✝ New York

To the memory of all my grandparents

Dedicated to my mom Zulema
To my dad Billy
To my brother Tulio
To my life brothers Federico and Allen

Thanks for the support

Everything should be made as simple as possible, but not simpler
(Albert Einstein)

CONTENTS

PREFACE

Biodiesel production is a very modern and technological area that is wining relevance and market due to its benefits, such as that is biodegradable, renewable and is an alternative source of fuel with less pollutants and less particle pollution compared to fossil fuels.

Different studies have been carried out using different oils as raw material (feedstock with different composition and purity), different alcohol as well as different catalysts: homogeneous ones, such as sodium hydroxide, potassium hydroxide, sulfuric acid and supercritical fluids, and heterogeneous ones; such as solid resins and enzymes as well as new technologies that are being developed every day.

Because of this, the need of a compilation of the ideas and work seems important as well as useful. With this goal, the purpose of this book is to develop some grounds and to establish the state of the art over some of all the possible technologies that are available to produce biodiesel. I would try to show the advantages of each of them as well as when one is more suitable to be used than other, but also the disadvantages. Point out the disadvantages could lead to new develops, optimization as well as improvements, being all of them the key of research.

This book will be separate into several chapters, in the first three chapters I will try to show you the global energy situation as well as the *today* of the biodiesel industry, what are the source, how is the market, etc. In the following chapters I will discuss several technologies individually, focusing in operational conditions, kinetics model and economic comparison in order to see if they could be use as profitable alternatives.

In the final chapter it will be summarized all the previous technical aspect of each alternative and a comparison will be done to point out the weakness and the strength of each process.

Chapter 1

GENERAL FUEL SITUATION

1. GENERAL WORLD OIL SITUATION

Derivative products of petroleum, such as fuel, as well as hydrocarbons for petrochemical industries were the most used compounds all over the world in the last century. Gasoline and diesel fuels have become the motor of the industrial world; their consumption has been increasing as well as the global demand has increased. Over the last years, global population has grown by over 6000 million inhabitants, demanding electricity, gas, fuels, etc. The increase in the world's population and its effect on domestic energy demands can be seen in Figure 1 [1].

This growth in the population and the industrialization of the world has made an increase of the petroleum consumption. Figure 2 shows the patterns of petroleum consumption of the last fifty years. It can be seen that this consumption has doubled since 1960 and it will continue growing until the oil peak might be reached or the price might be so high that consumption of petroleum fuel will decrease. However, if new reserves are found this tendency might continues.

This continuous augmentation in petroleum demands and consumption has a parallel effect on its price, producing an increase that in July-August 2008 had reached historical values of over 140 US$ per barrel of crude. Figure 3 shows an evolution of the average historical values in the last 12 years.

More than 85% of the total amount of energy supply comes from non renewable energy sources such as petroleum, carbon and gas. On the other hand, a small amount of energy is due to alternative renewable sources (in the case of biodiesel this amount is less than 1% [3]).

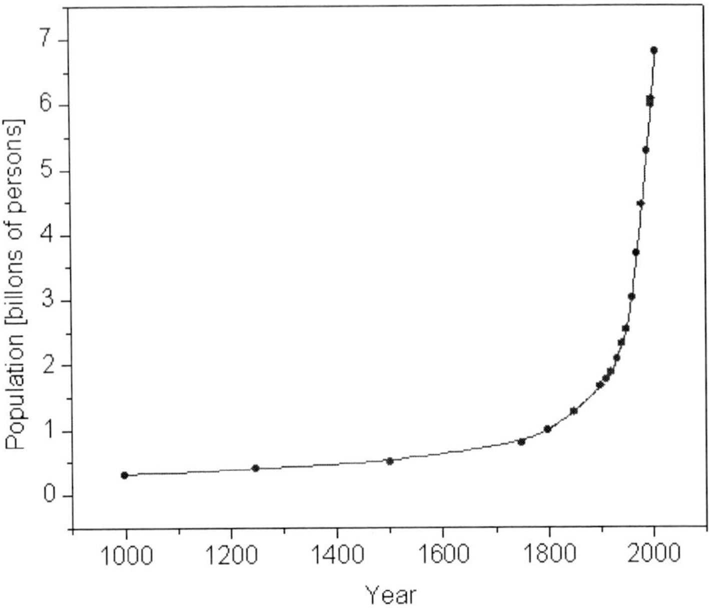

Figure 1. Evolution of the world population over the years [1].

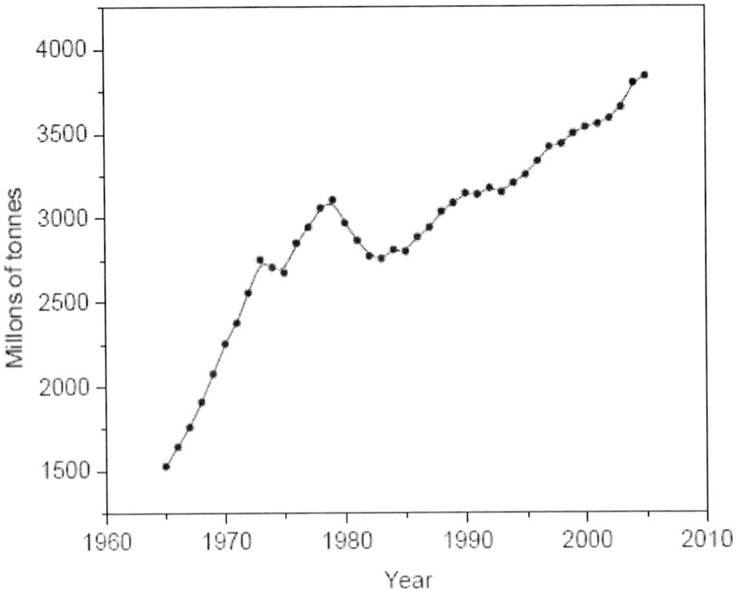

Figure 2. Consumption of petroleum fuel over the past 50 years [2].

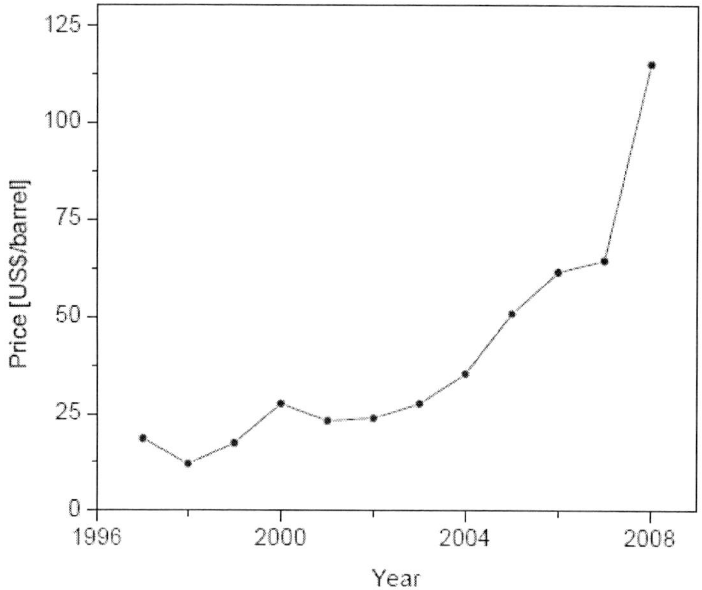

Figure 3. Variations of the price of crude oil. [4].

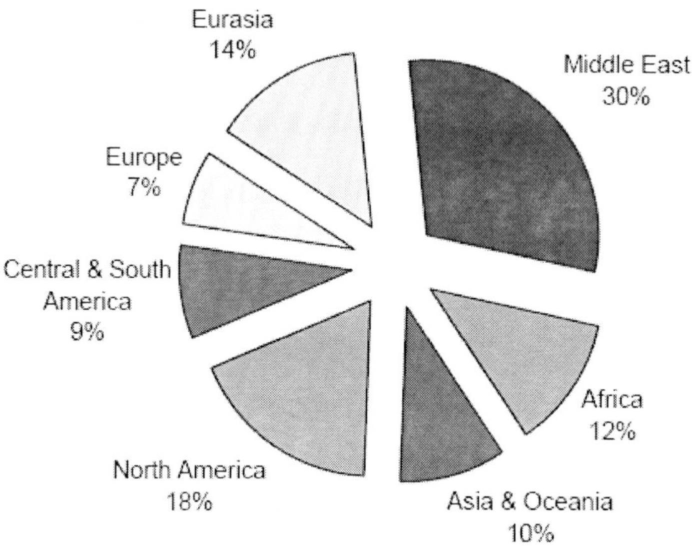

Figure 4. Production of oil. Percentage per region. Source EIA [6].

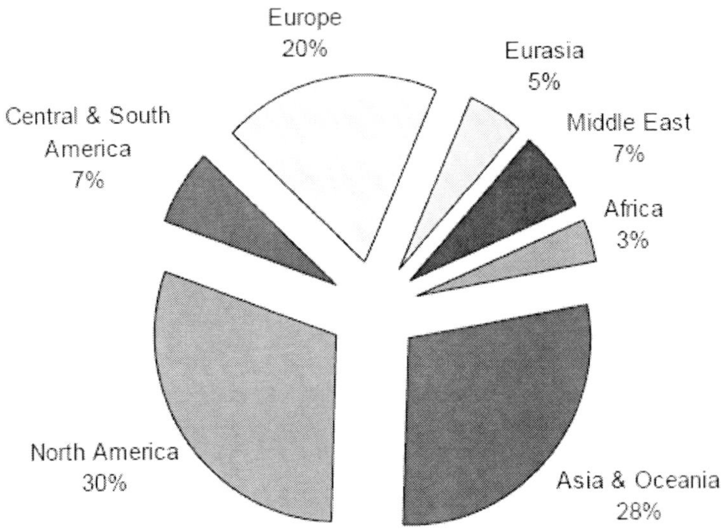

Figure 5. Oil consumption: Percentage per region. Source EIA [6].

In parallel with this continuous population growth, there is also an environmental issue fossil fuel emissions are a very important factor in the green house effect producing a reduction in the ozone layer and an increase in the global temperature [5]. These factors are the reason why alternative fuels, biodiesel, bioethanol, biogas, wind, solar energy, etc, continue to be developed with the aim of being less dependent on fossil fuel, and rely more on renewable and environmental friendly energy sources. Also, due to the vicinity to the Oil's peak[1] (in some cases it is believed that it has been already reached); biogas, bioethanol and biodiesel, among other alternatives fuels, will substitute part of the fossil fuel market.

The general oil production and consumption over the world, according to the International Energy Agency, is divided as showed in Figure 4 and Figure 5 [6]

Besides petroleum fuels, here are other major products generated in refineries. Several processes are being carried out simultaneously. According to the distillation and cracking properties and to the operative conditions several amounts of secondary products are produced. Although they are called secondary products their relevance to regular modern life is extremely high.

[1] Peak oil is defined as the situation where the extraction of petroleum reaches it maximums and from there the production will decreases in comparison to the demand.

Naphthanes, liquefied gas, bitumen, lubricants, are of great relevance in today's way of living.

2. Current Biofuel World Situation

Several alternative energies have been developed over the last years in order to use natural resources for fuel production. The main advantage is that they are more environmentally friendly compared to petroleum fuels [7-10]. Among these alternatives we can find bioethanol, biodiesel, biogas, wind, solar or water energy, hydrogen, etc [11-14].

Different countries have different alternative energy plans and projects. Several countries in the European Community are employing wind energy as a major source of energy. According to the World Wind Energy Association, USA, Germany and Spain are the leading countries with capacity installed for wind energy [15], while Norway produces mainly hydroelectric energy [16] even though it is not one of the largest producers (like China or Canada) of hydroelectric energy, its production satisfied more than 98% of the internal consumption. Germany and Japan lead the way in solar energy production [12] with around 80% of the world production, what place them between the top 15 countries that uses this alternative energy. Other sources such as hydrogen or biomass are also being studied and developed within the scope of finding a replacement for fossil derived fuels [11,17].

In order to replace fossil liquid fuel, bioethanol and biodiesel are becoming the most attractive alternative to substitute gasoline and petroleum diesel. General attention have been withdrawing to bioethanol produced from second generation raw materials (generally waste) [18-21].

This production could follow two alternative paths as proposed by Mielenza [22]. These two alternative ways are very different based on the process used before the sugar fermentation (See Figure 6.). One way employs enzymes to carry on the hydrolysis while a second alternative for this reaction is carried on with an acid, normally sulfuric acid. However, if the raw material used is considered a waste, some purification equipments before the hydrolysis might be required [23-26].

This liquid biofuel could be blend with gasoline and, at some point, to fully replace it. During 2007 Brazil and USA were the major producers of bioethanol with over 85% of the global production [14], while the total amount for 2005 was over 35000 millions liters.

The other major liquid fuel is biodiesel, which is defined as the mono alkyl ester derivate from lipid feedstock such as vegetable oils or animal fats (ASTM definition [27]). This fuel is generally produced from vegetable oils and, as it will be point out in Chapter 2, it has several environmental and technological advantages which make it the ideal substitute for petroleum diesel fuel.

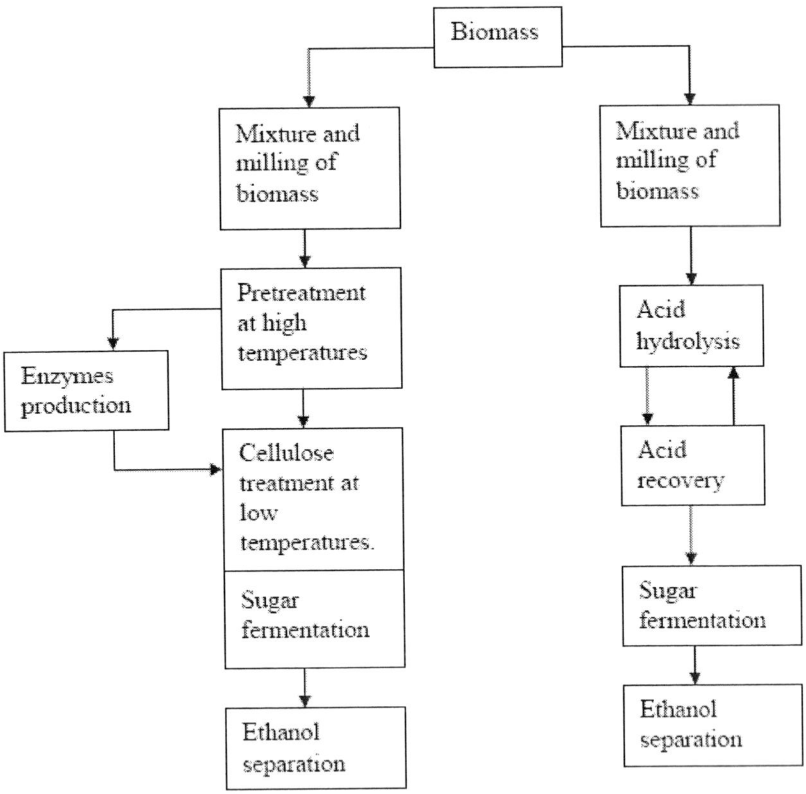

Figure 6. Production of bioethanol using biomass [22].

In the early 1900, Rudolf Diesel created the engine that was named after him and tried it with vegetable oils as fuel. At that time, oils were more expensive than petroleum and therefore, the engine was not used. His idea was storage for almost a century, until it has show up again as an alternative to mitigate the fuel crisis. Nevertheless, vegetable oils have not re-appeared as net fuels, but as raw materials which will be transformed into biodiesel. Biodiesel are of high quality for engines and they could be easily blended with regular diesel.

The first MAN (Maschinenfabrik Augsburg Nürnberg) diesel engine (Figure 7.) becomes operable using a mixture of lamp oil and gasoline and air pressure injected into the combustion chamber [28]. However, the new diesel engines use biodiesel fuel that has to satisfy international standards over several different aspects. In section 3 the required needs by the IRAM standards are shown.

Printed with permission.

Figure 7. Photo of the first diesel engine provided by MAN.

Table 1. EN 14214 standard for biodiesel quality control [29]

Required	Units	Limits	Experiments
Esters amount	% (mol/mol)	96.5	EN 14103
Density at 15°C	kg/m^3	860-900	EN ISO 3675, EN ISO 12185
Kinematics Viscosity at 40°C	mm^2/s	3.5-	EN ISO 3104, ISO 3105
Flash Point	°C	120 min	EN ISO 3679
Sulfur content	mg/kg	10.0 max	EN ISO 20846, EN ISO 20884
Carbon residue	% (mol/mol)	0.3 max	EN ISO 10370
Cetane Number		51 min	EN ISO 5165
Sulfated ash	% (mol/mol)	0.02 max	ISO 3987
Water content	mg/kg	500 max	EN ISO 12937
Total Contamination	mg/kg	24 max	EN 12662
Copper strip corrosion (3 h, 50 °C)	Degree of corrosion	1	EN ISO 2160
Oxidation stability, 110C	H	6.0 min	EN 14112
Acid number	mg KOH/g	0.50 max	EN 14104
Iodo number	g I_2/100 g	120 max	EN 14111
Linolenic acid content	% (mol/mol)	12.0 max	EN 14103
Polyunsaturated methyl ester	% (mol/mol)	1 max	EN 14103
Methanol content	% (mol/mol)	0.2 max	EN 14110
MAG content	% (mol/mol)	0.8 max	EN 14105
DAG content	% (mol/mol)	0.2 max	EN 14105
TAG content	% (mol/mol)	0.2 max	EN 14105
Free Glycerol	% (mol/mol)	0.02max	EN 14105, EN 14106
Total Glycerol	% (mol/mol)	0.25 max	EN 14105
Group I metals	mg/kg	5.0 max	EN 14108, EN 14109
Group II metal	mg/kg	5.0 max	EN 14538
Phosphorous content	mg/kg	10.0 max	EN 14107

3. Biodiesel Standard

Biodiesel is defined by ASTM [27] as the fuel made from vegetable oils or animal fats. However, this definition is wide and should be more specific because, otherwise, any blend of biodiesel-diesel is miss called biodiesel. The real name should be BX, where x stands for the percentage of biodiesel in the mixture. For example B20, is the most common blend is 20 % of biodiesel and 80% of regular diesel. However to have a fuel that might be call biodiesel, internationals standards should be satisfied. These standards vary from country to country even each region has its own policy as well. However, they are all quite similar and have similar values and restrictions. The most common ones are ASTM, DIN, and IRAM.

IRAM 6115-1, ASTM D6751-08, EN 14214 or DIN 51606 are some of the standards that should be attained in order to have a biodiesel fuel (B100). Table 1 shows a summary of the tests required, the maximum and /or minimum value of each variable, and the experiments to determine each quantity for the IRAM standard.

References

[1] http://www.un.org/esa/population/publications/sixbillion/sixbilpart1.pdf

[2] BP Statistical Review of World Energy June 2006.

[3] U.S. Department of Energy. July 31st 2006.

[4] http://eia.doe.gov

[5] http://www.epa.gov

[6] http://www.iea.org

[7] Marchetti. J.M.. Miguel. V.U.. Errazu. A.E. *Possible methods for biodiesel production*. Renewable and Sustainable Energy Reviews. 11. (2007), 1300-1311.

[8] Srivastava A. Prasad R.. *Triglycerides-based diesel fuels*. Renewable Sustainable Energy Reviews. 4. (2000), 111–133.

[9] Ma F. Hanna MA.. *Biodiesel production: A review*. Bioresource Technology. 70. (1999), 1–15.

[10] Fukuda. H., Kondo. A., Noda, H. *Biodiesel fuel production by transesterification of oils. Journal of Bioscience and Bioengineering.* 92(5). (2001), 405–416.

[11] http://www.eia.doe.gov/kids/energyfacts/sources/IntermediateHydrogen.html
[12] http://www.iea-pvps.org/isr/01.htm
[13] http://www.biodiesel.org
[14] http://www.ethanolrfa.org
[15] http://www.wwindea.org/home/images/stories/worldwindenergyreport2008_s.pdf
[16] http://www.economist.com
[17] http://www.eia.doe.gov/kids/energyfacts/sources/renewable/biomass.html
[18] Balata, M., Balat, H., Öz, C. *Progress in bioethanol processing.* Progress in Energy and Combustion Science. 34. (2008), 551-573.
[19] Kim, S., Dale, B.E. *Global potential bioethanol production from wasted crops and crop residues. Biomass and Bioenergy*. 26. (2004), 361-375.
[20] Rocha, M.V.P., Rodrigues, T.H.S., Macedo, G.R., Gonçalves, L.R.B. *Enzymatic Hydrolysis and Fermentation of Pretreated Cashew Apple Bagasse with Alkali and Diluted Sulfuric Acid for Bioethanol Production. Applied Biochemistry and Biotechnology*. 155(1-3). (2009), 104-114.
[21] Rabelo, S.C., Filho, R.M., Cost, A.C. *Lime Pretreatment of Sugarcane Bagasse for Bioethanol Production. Applied Biochemistry and Biotechnology.* 155(1-3). (2009), 139-150.
[22] Mielenza. J.R. *Ethanol production from biomass: technology and commercialization status. Current Opinión in Microbiology*. 4. 324-329. 2001.
[23] Chandra, R.P., Bura, R., Mabee, W.E., Berlin, A., Pan, X., Saddler, J.N. *Substrate Pretreatment: The Key to effective Enzymatic Hydrolysis of Lignocellulosics?. Advance Biochemistry Engineering and Biotechnology.* 108. (2007), 67-93.
[24] Rosgaard, L., Pedersen, S., Meyer, A.S. *Comparison of Different Pretreatment Strategies for Enzymatic Hydrolysis of Wheat and Barley Straw. Applied Biochemistry and Biotechnology*. 143. (2007), 284-296.
[25] Linde, M., Galbe, M., Zacchi, G. *Steam Pretreatment of Acid-Sprayed and Acid-Soaked Barley Straw for production of Ethanol. Applied Biochemistry and Biotechnology.* 129-132. (2006), 546-562.
[26] Yang, B., Wyman, C.E. *Pretreatment: the key to unlocking low-cost cellulosic ethanol. Biofuel, Bioproducts and Biorefining*. 2. (2008), 26-40.
[27] http://www.astm.org

[28] http://www.man.de

[29] Committee for Standardization Automotive fuels – fatty acid methyl esters (FAME) for diesel engines – requirements and test methods. European Committee for standardization, Brussels, 2003a. Method, 14214.

Chapter 2

GENERAL BACKGROUND ABOUT BIODIESEL

1. INTRODUCTION

Vegetable oils or animal fats are composed mostly of triglycerides, known as well as triacylglycerols (TG), where a fatty acid chain or several fatty acid chains are attached to glycerin molecules, as schematized in Figure 1. Fatty acid chains can have one double bond, two double bonds, three double bonds and some cases no double bonds at all. The composition of each vegetable oil depends on its precedence. This is well described by Ma and Hanna [1], for example, sunflower oil has fatty acid chain containing 18:2 and 18:1, while for crumble oil the majority is 22:1 and 16:1.

Biodiesel is produced through a reaction called transesterification, where the TG reacts with an alcohol, generally methanol in the presence of a catalyst to produce the biodiesel and a triol called glycerin. A typical reaction can be seen in Figure 2 [2]

This reaction is a three step reaction, where from triglycerides diglycerides are produced. From them monoglycerides are formed and finally these are transformed into glycerin. In each step biodiesel is produced [3-5]. This reaction can be observed in Figure 3.

As shown in Figure 4 diglycerides and monoglycerides are intermediate compounds, produced and consumed during the reaction. Therefore, they go through a maximum in their concentrations profile. Variations of the percentage of each compound could be seen in Figure 5.

Worldwide biodiesel production has been increasing over the last 20 years due to political decisions related to environmental matters, as well as economic considerations. The environmentally friendly biofuel development as well as the need of independence from petroleum economies have made the amount of biodiesel produced to grow considerably in the last years, from less than 750 millions of liters by year 2000 to around 4000 millions per liter by 2008 [7].

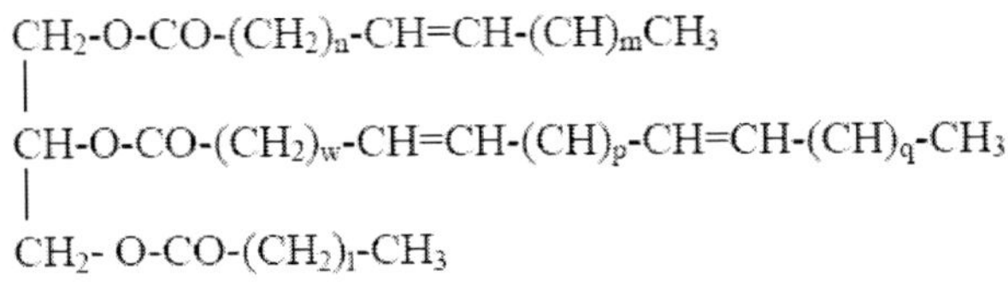

Figure 1. General representation of a triglycerides molecule. Where n, m, w, p, q, l, represents arbitrary length on the main chain.

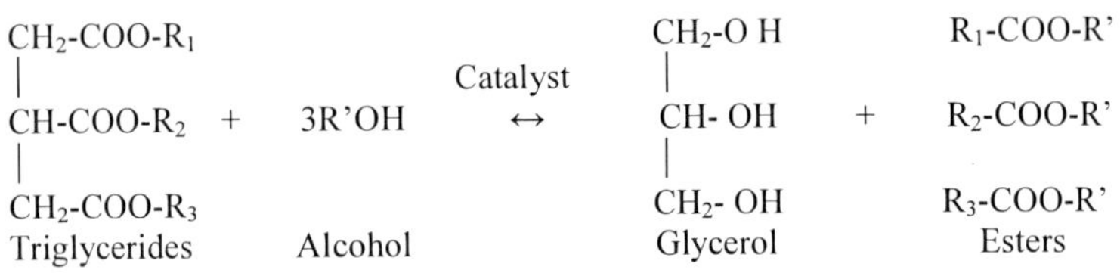

Figure 2. Typical transesterification reaction: a triglyceride with an alcohol to produce biodiesel. R_n are carbon chain of different length.

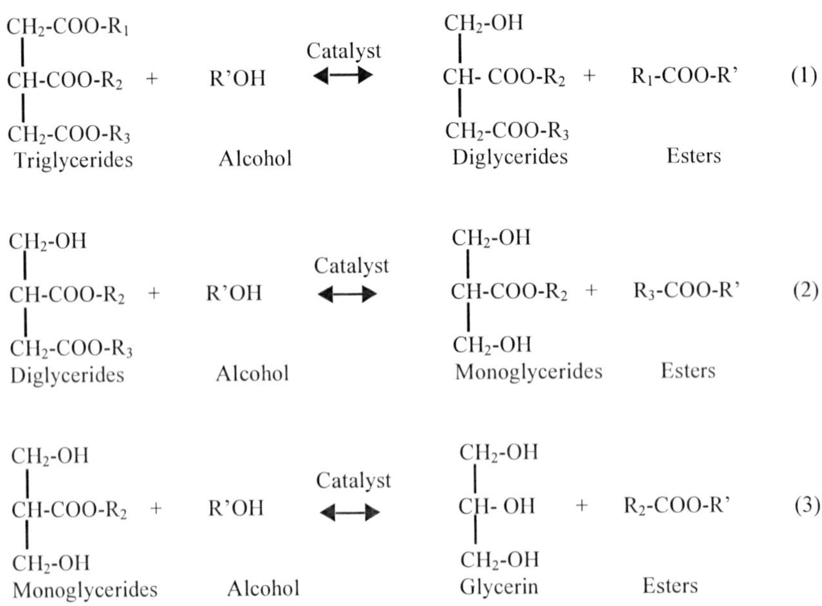

Figure 3. Series of reactions involved in the transesterification reaction.

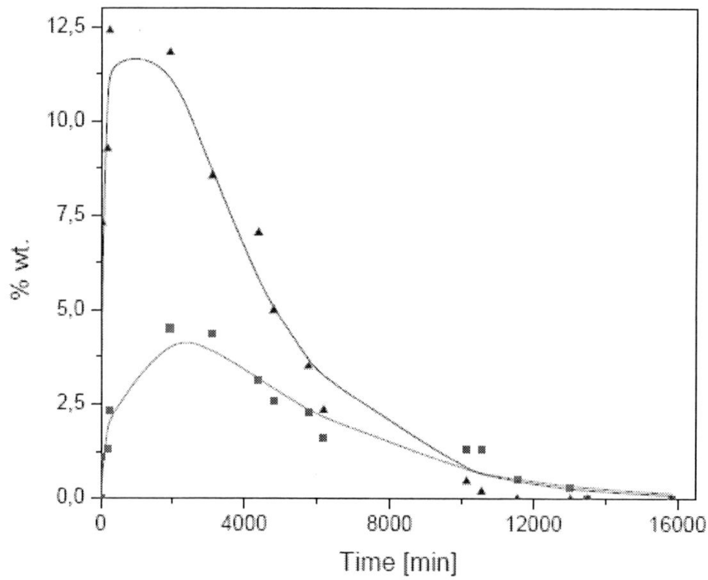

Figure 4. Weight percentage vs. time. (▲) diglycerides, (■) monoglycerides [6].

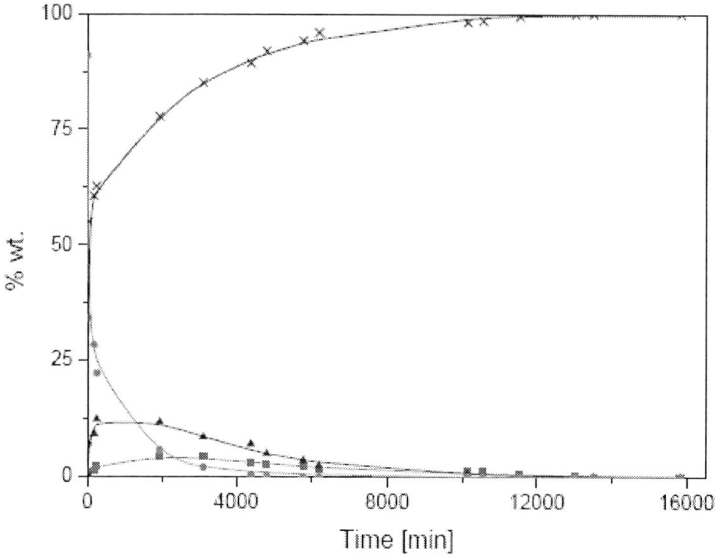

Figure 5. Weight percentage vs. time. (●) triglycerides, (▲) diglycerides, (■) monoglycerides, (x) esters [6].

Some of the main advantages and disadvantages [6] that have been encouraging this increase in the biodiesel production are enumerated as follows:

1.1. Advantages

- It is biodegradable compared with petroleum diesel fuel. This is because of its biological properties and origin.
- It achieves a better combustion due to the presence of an oxygen molecule in the ester carbon chain.
- It has almost no CO emissions due to the better combustion of the fuel.
- It has a higher CO_2 reduction emission, over 45 % and in some cases over 74% compared to petroleum diesel emissions.
- There is not sulfur compound emission from its combustion, or if there are some, they are highly reduced.
- Particle pollutants are extremely reduced.
- A cutback on the ashes produced from engines is achieved.
- It has a better lubricant effect over the engine due to the oil properties of the raw material.
- Engines' shelf life is highly increased due to the fuel properties, as mention, high lubricity.
- It is easier and safer to transport, to handle and to storage, compared with diesel fuel provided from petroleum.

1.2. Disadvantages

- The freezing temperature of biodiesel or biodiesel-petroleum diesel blends is lower compared to fossil fuel. This produces the need of including additives in biofuel in order to employ them in cold weather, what makes it less natural.
- Car filters are more likely to be dirty and need to be changed due to the use of a biodiesel-diesel mixture-
- Biodiesel could not be storage for long periods of time due to its biodegradability, producing a loss of its properties.

- Increasing on the NO_X emissions of biodiesel engines have been measured.
- Due to the fact that the raw material used is regular oil, a competition between oil for fuel and oil for food has arrived. Laws that make the use of biodiesel an obligation are also making the price of feed oil to increase.

As it will be point out in the following chapters, several catalytic technologies are being employed, or under research, to carry on the transesterification reaction, such as:

- Homogeneous basic catalysis: sodium or potassium hydroxide.
- Homogeneous acid catalysis: sulfuric acid is generally used
- Enzymes catalysis
- Heterogeneous catalysis: solid resin (acid or basic) as well as support oxides such as calcium oxide over alumina.
- Supercritical reaction: this process does not require a catalyst
- In situ transesterification.

Each technology implies a different layout and different types of equipment requirements to produce the biofuel. This is due to the fact that each process has different steps, normally after the transesterification section. However, if it is required, combinations of two technologies can be easily done in order to treat several raw materials and improve the global efficiency of the plant.

2. General Considerations of the Biodiesel Production Process

A conventional process flow diagram is summarized in Figure 6. For a conventional homogeneous basic catalyst process where the reaction is followed by separation. The most common purification steps are water washing follows by distillation and separation of the final product. In some other cases it could be used water washing followed by catalyst neutralization and continue with other purifications such as regular separation or distillation [8-11].

The goal of each of these alternatives is to provide a final product that satisfies the international standards as seen in section 3. Purification for the secondary product, glycerin, is required in order to sell it in the industrial market.

If a heterogeneous catalyst is employed, or a supercritical process is considered, then the schematic flow sheet is much simple and requires fewer amounts of equipments. Figure 7 shows a block diagram for a heterogeneous catalyst, where it can be seen that the separation of the catalyst and the purification of the biodiesel as well as of the glycerin are considered together. This is possible due to the fact that no external streams are required to obtain a biodiesel that satisfy international standards.

As it will be seen in Chapter 3, if the amount of free fatty acid is over 0.1-0.5% wt. [1,3,12], the conventional basic homogeneous catalytic process is not suitable to carry on the reaction due to the high amount of soap formation. This undesirable product makes the down streaming separation complicated and expensive. However, depending of the catalyst and the process employed, several possible side reactions might take place simultaneously with the transesterification reaction.

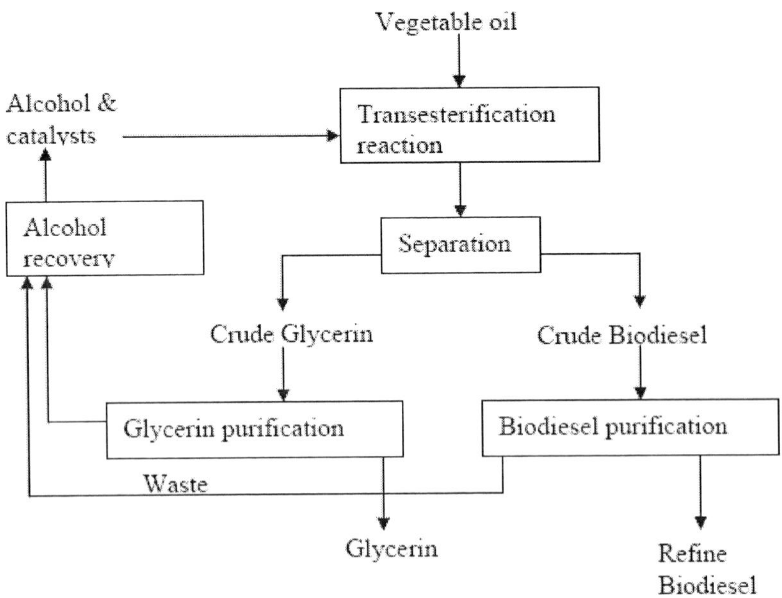

Figure 6. Basic process flow sheet for a conventional production process of biodiesel [12].

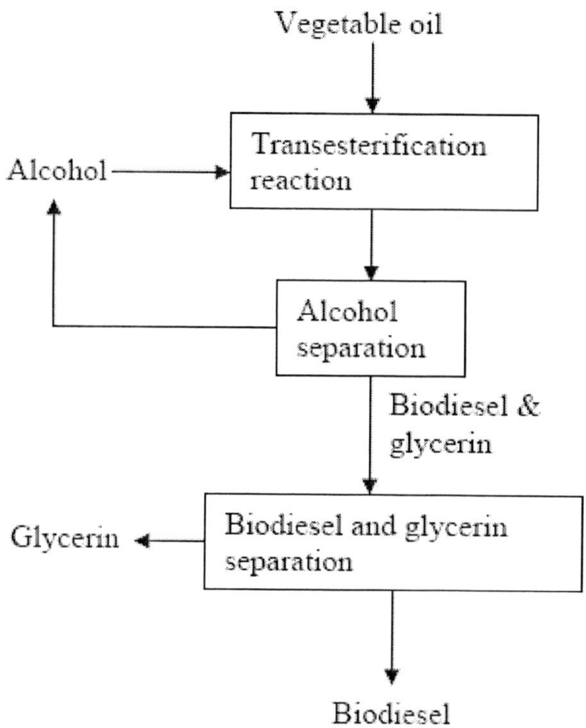

Figure 7. Flow diagram of biodiesel production using a heterogeneous catalyst [6].

If a basic homogeneous catalyst is used, the most common and fast side reaction that will be happening is the soap formation one:

R-COOH	+	NaOH	$\longleftrightarrow$	H_2O	+	R-COO-Na	(4)
Fatty acid		Base		Water		Soap	

While if the catalyst is sulfuric acid or a solid resin, the most common side reaction will be the esterification of the free fatty acid to produce esters.

R-COOH	+	R_2OH	$\overset{\text{Catalyst}}{\longleftrightarrow}$	H_2O	+	R-COO-R_2	(5)
Fatty acid		Alcohol		Water		Ester	

In the later, the production of the biofuel it is increase due to a compound that is generally considered an impurity in the biodiesel industry. A brief comparison of the major variables involved in several technological alternatives has been done and it is showed in Table 1

Table 1. Comparison of different technologies in the biodiesel production industry [13]

Variable	Base	Enzyme	Supercritical	Monolithic	Resin	Acid
Temp. [°C]	60-70	30-50	200-350	50-180	60-180	50-80
FFA	Soaps	Esters	Esters	Esters	Esters	Esters
Ester yield	Normal	High	High	Normal	Good	Normal
Purity of glycerin	Difficult	Simple	Simple	Simple	Simple	Difficult
Reaction time	1-2 h	8 h	4 min	6 h	Variable	4 h -3 days.
Ester purification	Difficult	Simple	Simple	Simple	Simple	Difficult
Cost	Cheapest	Expensive	Expensive	Medium	Medium	Cheaper
Amount of equipment	High	Low	Low	Low	Low	High

Table 1 it is just presented here, it will be in depth discussed (Chapter 10) when all the technologies have been discussed to a more detail level, following chapters.

References

[1] Ma F. Hanna MA.. *Biodiesel production: A review. Bioresource Technology.* 70. (1999), 1–15.

[2] Marchetti. J.M.. Miguel. V.U.. Errazu. A.E. *Possible methods for biodiesel production. Renewable and Sustainable Energy Reviews.* 11. (2007), 1300-1311.

[3] Srivastava A. Prasad R. *Triglycerides-based diesel fuels. Renewable Sustainable Energy Reviews.* 4. (2000), 111–133.

[4] Fukuda. H., Kondo. A., Noda, H. *Biodiesel fuel production by transesterification of oils. Journal of Bioscience and Bioengineering.* 92(5).(2001) 405–416.

[5] Knothe, G., Van Gerpen, J., Krahl, J. (Eds.), *The biodiesel handbook*, AOCS Press, Champaign, Illinois, 2005, 302 pp

[6] Marchetti, J.M. Ph.D. Thesis. *Technological Alternatives for Biodiesel Production.* Universidad Nacional del Sur, Bahía Blanca, Argentina, 2008.

[7] http://www.iea.gov (International energy agency)

[8] Marchetti, J.M., Errazu, A.F. *Technoeconomic study of supercritical biodiesel production plant. Energy Conversion and Management.* 49. (2008), 2160-2164.

[9] Haas, M.J., McAloon, A.J., Yee, W.C., Foglia, T.A. A *process model to estimate biodiesel production costs. Bioresource Technology.* 97(4). (2006), 671-678.

[10] Van Gerpen, J. *Biodiesel processing and production. Fuel Processing Technology.* 86, (2005), 1097-1107.

[11] Zhang Y, Dubé MA, McLean DD, Kates M. *Biodiesel production from waste cooking oil: 1. Process design and technological assessment. Bioresource Technology.* 89. (2003), 1–16.

[12] Barnwal, B.K. Sharma, M.P. *Prospects of Biodiesel production from vegetable oils in India. Renewable and Sustainable Energy Reviews.* 9(4). (2005) 363-378

[13] Marchetti, J.M. *Past, Present and Future Scopes in the Biodiesel Industry.* World Conference of Science, Engineering and Technology. Oslo, 2009.

Chapter 3

ALTERNATIVE FEEDSTOCKS

1. INTRODUCTION

Biodiesel is produced, as previously mentioned, from vegetable oils or animal fats. However, which oil will be employed it is directly related to the kind of process that will be used (see following chapters) but also with the availability and price of different oil options in the region where the biodiesel plant should be installed. A good summary of several mayor plantations and where they have the larger cultivated areas was showed by Obrien et al. [1]. Table 1 is extracted from their publications [1]. It could be seen that depending on the seed, which countries have the mayor productive areas.

Despite the fact that some countries have a much higher production from one type of plant that others, due to climate and to internal policies, each grain or seed has a different amount of oil to be extract and therefore, a different amount of biodiesel that could be produce. This is not related to the production from tonne of vegetable oil but to the production of biodiesel from each plant. Figure 1 shows the general production of biodiesel from a sunflower seed.

Even more, each plant has its productivity for hectare planted. Some of them are showed in Table 2 where it can be seen that there are a very with range depending on the vegetable that wants to be cultivate.

Because of all this, it can be point out that there are many variables involved regarding the best selection to what grain should be used to obtain the mayor productivity. Not only performance should be considered, but also proximity, availability, production, lands, market for the final product, etc. Showing that the choice is complicate and need careful attention.

Table 1. Mayor producer for several vegetable oils [1]

Seed	Amount of oil (%)	Productive areas
Canola	40-45	Canada, China, India, France, Austria, United Kingdom, Germany, Poland, Denmark, Check Republic.
Corn	3.1-5.7	USA, Mexico, Russia, Belgium, France, Italy, Germany, Spain, United Kingdom.
Cotton	18-20	China, Russia, USA, India, Pakistan, Brazil, Egypt, Turkey.
Peanut	45-50	China, India, Nigeria, USA, Senegal, South Africa, Argentina
Crocus	30-35	China, USA, Spain, Portugal
Soybean	18-20	USA, Brazil, Argentina, China, India, Paraguay, Bolivia
Sunflower	35-45	Russia, Argentina, Austria, France, Italia, Germany, Spain, United Kingdom.
Coconut	65-68	Filipinas, Indonesia, India, México, Sri Lanka, Thailand, Malaysia, Vietnam, Mozambique, New Guinea, Republic of Côte d'Ivoire
Olive	15-35	Spain, Italy, Italia, Greece, Tunes, Turkey, Morocco, Portugal, Syria, Algeria, Yugoslavia, Egypt, Israel, Libya, Jordan, Lebanon, Argentina, Chile, Mexico, Peru, USA, Australia.
Palm	45-50	Malaysia, Indonesia, China, Filipinas, Pakistan, México, Bangladesh, Colombia, Nigeria, Republic of Côte d'Ivoire
Palm kernel	44-53	Malaysia, Indonesia, China, Filipinas, Pakistan, México, Bangladesh, Colombia, Nigeria, Republic of Côte d'Ivoire

Although all these variables, there are two mayor drawbacks in the biodiesel production related to the raw oil.

1) The high price of refine oil
2) The possibility to be using a food for fuel instead a food to feed.

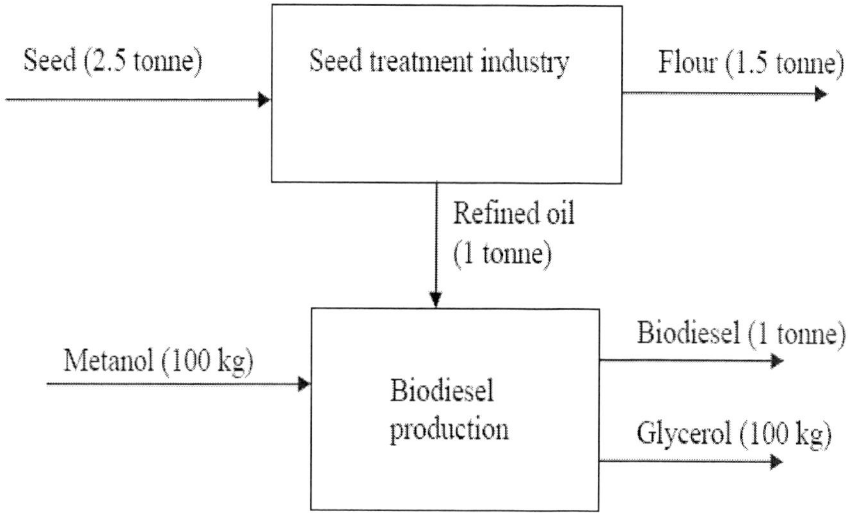

Figure 1. Process for biodiesel production from crop seed.

These two mayor concerns are now a days the engine to the search of new alternative, cheaper and sustainable raw materials for biodiesel production. High impure raw materials, normally known as second or third generations vegetable oil, are gaining more and more relevance in order to be use to produce biodiesel. This high impure oil have a much lower price than refine oil, refines oil could be around 500 US$ per tonne [2], while frying oil are around 110-220 US$ per tonne [2-3] (more discussion at the end of the chapter). Even more, the frying, cooking or waste oil, as its name indicated, are contaminants to the environment, normally polluting drinking water. The use of them for biodiesel production provides three main advantages

1) A much lower price raw material.
2) Raw oil that could not be use for food purposes, avoiding the use of refines oil that could be use as food.
3) The treatment of a pollutant that could contaminate high amounts of drinking watcr.

In the following section, a more detailed description of the possible vegetable oil use for biodiesel production will be introduced.

2. Types of Raw Vegetable Oil

This fuel it is generally produced from refine oil, defined as oil that has been degummed, with no phospholipids, tocopherols, water, and free fatty acid among other properties. This type of oil it is the most common due to the most used process to produce. Nevertheless, there are some disadvantages in using refine oil, as mentioned before. Demirbas [4] presents a list of the properties of several oils; in Table 2 it is showed the average of those values for each property. It was also added some range for crude oil definition as well as waste, frying or cooking oil.

As it can be seen the amount of impurities is considerable low for refine oil due to the fact that it might be used for human consumption. While crude oil, which might still be healthy and could be eaten, it allows a much higher amount of impurities. In the case of frying oil, other compounds have a much higher percentage, making it normally a waste and therefore a pollutant for the environment.

Table 2. Properties of the mayor types of oil

Property	Refine Oil [4-5]	Crude Oil [6-7]	Waste Oil [2, 8-10]
Kinematic viscosity [mm^2/s]	30.2	36	40.2
Carbon residue [wt%.]	0.24	0.278	0.18
Cetane number	38.01	39	---
Higher heating value [MJ/kg]	39.41	39.2	24.67
Ash content [wt%]	0.012	0.01	---
Sulfur content [wt%]	0.013	0.075	5
Iodine value [cg I/g Oil]	112.86	125	13.2
Acid value [mg KOH/g Oil]	<0.2	variable	5.96

Another important source of raw material is non edible oils; in this case no problems between food and fuel will be encounter due to the fact of non edibility of some vegetable oils [5-6]. Table 3 shows some comparison between edible and non edible oils related to the mayor parameters of the biodiesel produces from these sources.

Table 3. Properties of biodiesel from different source

Parameters	Non Edible Oils				Edible Oils		
	Jatropha	Rubber	Castor	Pongamia pinnata	Soybean	Palm	Rapeseed
Viscosity	4.8	5,81	---	4.8	4.50	4.42	4.08
Specific gravity	---	0.874	0.960	---	0.882	0.86	0.885
Calorific value [MJ/kg]	39.23	36.50	39.5	---	37.00	---	39.76
Flash point [°C]	135	130	260	150	170	182	69
Cloud point [°C]	---	4	-12	---	-4	15	-2
Pour point [°C]	2	---	-32	---	-12	15	-3
Ash content [wt%]	0.012	...	0.020	0.0005	...	0.020	---
Acid value [mg KOH/g]	0.400	0.118	---	0.620	---	0,080	---

Extracted from reference [11] and references in within.

Table 4. Vegetable oils composition

Vegetable oil	Fatty acid composition % by weight								
	16:1	18:0	20:0	22:0	24:0	18:1	22:1	18:2	18:3
Corn	11.67	1.85	0.24	0.00	0.00	25.16	0.00	60.60	0.48
Cottonseed	28.33	0.89	0.00	0.00	0.00	13.27	0.00	57.51	0.00
Crambe	20.7	0.70	2.09	0.80	1.12	18.86	58.51	9.00	6.85
Peanut	11.38	2.39	1.32	2.52	1.23	48.28	0.00	31.95	0.93
Rapeseed	3.49	0.85	0.00	0.00	0.00	64.4	0.00	22.30	8.23
Soybean	11.75	3.15	0.00	0.00	0.00	23.26	0.00	55.53	6.31
Sunflower	6.08	3.26	0.00	0.00	0.00	16.93	0.00	73.73	0.00

Extracted from reference [1].

Edible oil as well as non-edible oils does not have the same main composition. Depending on the vegetable oil that it is used, the percentage and distribution of the triglycerides fatty acid chains, normally based on C18:1 and C18:2. Table 4 shows the mayor composition of fatty acid for several types of oil.

3. Newest Raw Material

Over the last decades, the search for new source of raw materials for biodiesel production has encourage the used of crude oil, soapstocks, waste oil, frying oil, etc. This search for new alternatives has produced new source of oil to be investigated and new alternatives have appeared on the market: oil coming from algae.

Over the last year, the consequences of using a refine vegetable oil for fuel purposes have raise the conflict between the need of food for the humans as well as the need of alternative energies. As previously mention, this conflict has led to the use of waste, crude oil, soapstocks as well as non edible oil. However, new options have common along, being the production of oil from algae and microalgae the new source.

Algae are very simple organisms that are aquatic and in the microscopic scale. They can be used for several types of renewable fuels, such as methane production from anaerobic digestion [12], photo biologically produced of biohydrogen [13] and biodiesel from microalgae oil [14]. The used of microalgae for biofuel has several advantages [15]:

- Removal of CO_2 from industrial gases and engine combustion. It will consume CO_2, decreasing gases from the green house effects while producing a biofuel.
- Wastewater treatment. Algae could easily grow in water that is not suitable for human consumption.
- The waste from the algae after being used for oil production could still be used for other chemical compounds, such as bioethanol.
- Algae could be easily produced in areas where no other agricultural seed might grow.
- Several other chemical compounds could be produced from algae as feedstocks.

Demirbas [16] presented a chemical composition of several algae, showing the main contribution of lipids, carbohydrates, proteins and nucleic acid. Nevertheless, if the conditions are favorable, algae produces oil, being this oil content variable according to the source. In Table 5 it is presented several microalgae sources and they oil content. This table was extracted from Chisti Y. work, reference [17].

Table 5. Oil content for several microalgae [17]

Microalgae	Oil content (% dry wt)
Botryococcus braunii	25-75
Chlorella sp.	28-32
Crypthecodinium cohnii	20
Cylindrotheca sp.	16-37
Dunaliella primolecta	23
Isochrysis sp.	25-33
Monallanthus salina	>20
Nannochloris sp.	20-35
Nannochloropsis sp.	31-68
Neochloris oleoabundans	35-54
Nitzschia sp.	45-47
Phaeodactylum tricornutum	20-30
Schizochytrium sp.	50-77
Tetraselmis sueica	15-23

Table 6. Oil content for several microalgae [15]

Plant source	Biodiesel productivity [kg biodiesel/ha year]
Corn	152
Hemp	321
Soybean	562
Jatropha	656
Camelia	809
Canola	862
Sunflower	946
Castor	1,156
Palm oil	4,747
Microalgae (low oil content)	51,927
Microalgae (medium oil content)	86,515
Microalgae (high oil content)	121,104

Chisti [17] also compared the production of oil from algae in terms or required landscape. The author has found that the production of oil is at least one order of magnitude higher than the best vegetable oil from crop. And in some cases it could be over 23 times bigger. The author have also showed that the required amount of hectares needed to satisfied 50% of the market varies from 1540 M ha for corn to 2 M ha when using a microalgae with 70% of oil in biomass. The most common amount of oil per dry weight of biomass is between 20 and 50% [17], in these scenario the required landscape is 4.5 M ha, being 10 times smaller than the required land area if palm oil were produced for biofuel purposes. This presents a new and very valuable source of oil for the biodiesel industry. Even more, Mata et al. [15] have done a comparison on the biodiesel productivity for several plant sources (Table 6. extracted from reference [15])

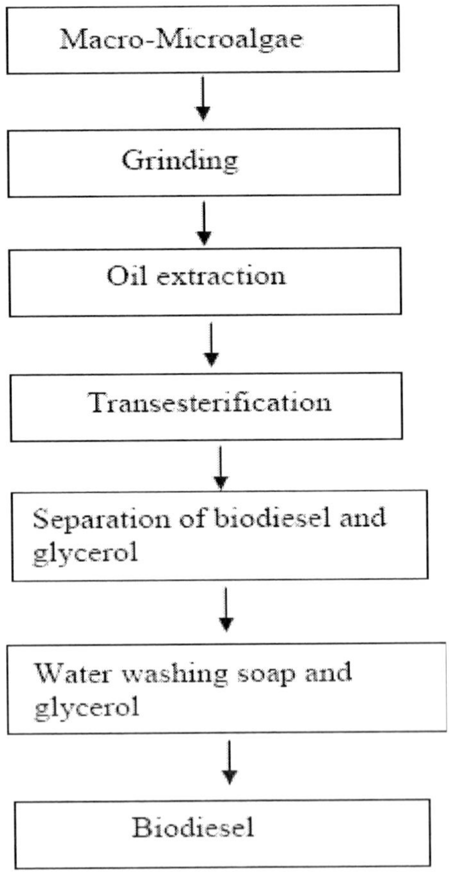

Figure 2. Production of biodiesel from Algae.

In order to produce this microalgae oil, microalgae biomass is required. This last compound production is based on sun light, carbon dioxide, water and some inorganic salts. The sunlight requirement is one of the most relevant due to the fact that during night, 25% of the biomass produced might be lost due to respiration [17]. Nevertheless, the consumption of CO_2 from microalgae is in a ratio close to 2 ton of CO_2 consumed per 1 ton of algal biomass, showing a great advantage for this technology and this new raw material.

In order to establish how this new raw material for biodiesel production could be used in a biodiesel plant, Demirbas [16] have proposed a schematic block diagram of the process form microalgae to final biodiesel, this is showed in Figure 2 (extracted from reference [16]).

Table 7. Properties of different biodiesel

FAEE	Melting point (°C)	$\Delta_c H$ (MJ/mol)	Kinematic viscosity (mm^2/s)	Oil stability index (h)	Cetane Number	Lub (μm)
C12:0 ME	5	8.14	2.43	>40	67	416
C12:0 EE	-2		2.63	>40		
C14:0 ME	19	10.67	3.30	>40		353
C14:0 EE	12		3.52	>40		
C16:0 ME	31	10.67	4.38	>40	86	357
C16:0 EE	19		4.57		93	
C16:1 ME	-34	10.55	3.67	2.1	51	246
C16:1 EE	-37					
C18:0 ME	39	11.96	5.85	>40	101	322
C18:0 EE	32		5.92	>40	97	
C18:0 BE	28		7.59		92	
C18:1 ME	-20	11.89	4.51	2.5	59	290
C18:1 EE	-20		4.78	3.5	68	
C18:1 BE	-26		5.69		62	303
C18:2 ME	-35	11.69	3.65	1.0	38	236
C18:2 EE			4.25	1.1	40	
C18:3 ME	-52	11.51	3.14	0.2	23	183
C18:3 EE			3.42	0.2	27	

Extracted from reference [18].

Table 8. Price of different raw materials

Raw bulk source	Price (US$/Tonne)	Reference
Waste Oil	110	2
Waste Cooking Oil	200	19
Waste Oil	220	3
Tallow	240	20
Hydrolyzed Feather meal	250	20
Grease	320	20
Blood	400	20
Acid Oil	400	21-22
Edible Palm Oil	478	11
Virgin Soybean oil	487	23
Pure Canola Oil	500	19
Edible Rapeseed Oil	683	11
Edible Soybean Oil	684	11

4. Biodiesel Properties from Different Vegetable Oils

Although the fact that biodiesel is produces from vegetable oil, the composition of the oil, such as the fatty ester composition, dictates what types of properties de final product will have. Even more, the presence of impurities or minor amounts of undesirable products will modified the properties of the final fuel.

The main variables that will be affected by this different composition are low temperature operability, oxidative and storage stability, kinematics viscosity, exhaust emissions, cetane number, and energy content [18]. Moser [18] presented a table comparing several properties of different final biodiesel. He has produced the fuel by using different carbon length vegetable oils and also using mainly methanol or ethanol. He also used different degree of saturation on the fatty acid chains, using double as well as triple double bonds in the carbon chain.

From Table 7 extracted from reference [18], it could be seen that the Oil stability index is very low for large hydrocarbons chains, such as C18:3, being over 40 for C12 to C16, all of them having no double bonds. Regarding the cetane number it is found that in almost all cases the cetane number is over the limit imposed by international standards (47 for ASTM D6751) there one few cases C18:2 and C18:3 where the cetane number was bellow the limit. From Table 7 it can be seen that the heat of combustion increases as well as the carbon chain increases.

Last but not least important, and not too much related to the properties of the biodiesel, we should not forget the economic factor: the price of the raw material. As it will be shown throughout the subsequent chapters of the book, the price of the raw material has a very important effect on the biodiesel industry. Here we will just compared some prices related to refine oil, waste oils, animals fats, etc, as presented in Table 8.

As it could be seen from Table 8. price increases as the purity of the raw material increases; however, for similar quality oils, price also are variable according to the main crop where the oil is produced, this is related to the need of processing this crop to produce oil.

References

[1] O'Brien, R., Farr, W., Wan, P. Editors. *Introduction to Fats and Oils Technology* Editores: Second 2000. Editorial AOCS.

[2] Zhang, Y., Dubé, M.A., McLean, D.D., Kates, M. *Biodiesel production from waste cooking oil: 2 Economic assessment and sensitivity analysis. Bioresource Technology*. 90. (2003), 229-240.

[3] Haas, M.J., McAloon, A.J., Yee, W.C., Foglia, T.A. *A process model to estimate biodiesel production costs. Bioresource Technology*. 97(4). (2006), 671-678.

[4] Demirbas, A. *Biodiesel. A Realistic Fuel Alternative for Diesel Engines*. Editorial Springer 2008.

[5] Chhetri, A.B., Watts, K.C., Islam, M.R. *Waste Cooking Oil as an Alternate Feedstock for Biodiesel Production. Energies*. 1. (2008), 3-18.

[6] Ion, O., Lucian, M., Panait, T., Adelina, P., Gheorghe, R. *Vegetable oil - renewable fuel for electricity and heat generation*.

[7] http://globalsmartinvestment.com/SoyOil.aspx

[8] Dmytryshyn, S.L., Dalai, A.K., Chaudhari, S.T., Mishra, H.K., Reaney, M.J. Synthesis and Characterization of Vegetable Oil Derived Esters: Evaluation for their Diesel Additive Properties. *Bioresource Technology*. 92. (2004), 55.64.

[9] Phan, A.N., Phan, T.M. *Biodiesel Production from Waste Cooking Oils*. Fuel, 87. (2008), 3490-3496.

[10] Anastopoulos, G., Zannikou, Y., Stournas, S., Kalligeros, S. *Transesterification of Vegetable Oils with Ethanol and Characterization of the Key Fuel Properties of Ethyl Esters*. *Energies*. 2. (2009), 362-376.

[11] Gui, M.M., Lee, K.T., Bhatia, S. *Feasibility of Edible oil vs. non-edible oil vs. waste edible oil as biodiesel feedstocks*. *Energy*. 33. (2008), 1646-1653.

[12] Spolaore, P., Joannis-Cassan, C., Duran, E., Isambert, A. *Commercial applications of microalgae*. *Journal of Bioscience and Bioengineering*. 101. (2006), 87-96.

[13] Ghirardi, M.L., Zhang, J.P., Lee, J.W., Flynn, T., Seibert, M., Greenbaum, E., Melis, A. *Microalgae: a green source of renewable H_2*. *Trends Biotechnology*. 18. (2000), 506-511.

[14] Dunahay, T.G., Jarvis, E.E., Dais, S.S. Roessler, P.G. *Manipulation of microalgal lipid production using genetic engineering*. *Applied Biochemistry and Biotechnology*. 57-58. (1996), 223-231.

[15] Mata, T.M., Martins, A.A., Caetano, N.S. *Microalgae for biodiesel production and other applications: A review*. *Renewable and Sustainable Energy Reviews*. 14. (2010), 217-232.

[16] Demirbas, A. *Production of Biodiesel from Algae Oils*. Energy Sources, Part A: Recovery, Utilizations and Environmental Effects. 31:2. (2009), 163-168.

[17] Chisti, Y. *Biodiesel from microalgae*. *Biotechnology Advances*. 25. (2007), 294-306.

[18] Moser, B.R. *Biodiesel production, properties and feedstocks*. In Vitro Cellular and Developmental Biology – Plants. 45. (2009), 229-266.

[19] West, A.H., Posarac, D., Ellis, N. *Assessmen of four biodiesel production processes using HYSYS.Plant*. *Bioresource Technology*. 99. (2008), 6587-6601.

[20] Canakci, M. *The potential of restaurant waste lipids as biodiesel feedstocks*. *Bioresource Technology*. 98. (2007), 183-190.

[21] Marchetti, J.M., Errazu, A.F. *Technoeconomic study of supercritical biodiesel production plant*. *Energy Conversion and Management*. 49(8). (2008), 2160-2164.

[22] Marchetti, J.M., Miguel, V.U., Errazu, A.F *Techno-economic study of different alternatives for biodiesel production*. *Fuel Processing Technology*. 89(8). (2008), 740-748.

[23] You, Y.D., Shie, J.L., Chang, C.Y., Huang, S.H., Pai, C.Y., Yu, Y.H., Chang, C.H. *Economic Cost Analysis of Biodiesel Production: Case in Soybean oil*. *Energy and Fuels*. 22(1). (2008), 182-189.

Chapter 4

HOMOGENEOUS BASE TECHNOLOGY

1. INTRODUCTION

Basic homogeneous technology is the most common and one of the most studied alternatives to convert vegetable oil into biodiesel. It is a good and reliable process that it is carried on under regular operational conditions.

The two more used catalysts, since there are several homogeneous alternatives are sodium hydroxide and sodium methoxide. Normally, the alcohol and the base are mix together in order to produce the alcohoxide, but sometimes it is not necessarily that way [1-7].

Methanol is the alcohol of preference due to its lower price; however, it is more than possible to used ethanol and obtained good results. Sometimes potassium hydroxide it is used, but less frequently that sodium hydroxide [1-8].

The reaction taking place for the formation of the alcohoxide, when methanol is used, is the following one [5]:

$$CH_3OH + Na(OH) \leftrightarrow CH_3O\text{-}Na + H_2O \qquad (1)$$

This reaction, if it takes place in the process, it is normally carried out separately from the transesterification reaction. This is done in order to obtain a more pure catalyst and to avoid the possibility of interaction of each of them separately with the vegetable oil producing the same final product but with a possible larger reaction time. A typical description of the process could be seen in Figure 1, where it can be observed that in the equipment label as PRE, the base catalyst and the alcohol are fed in order to produce the alkoxyde.

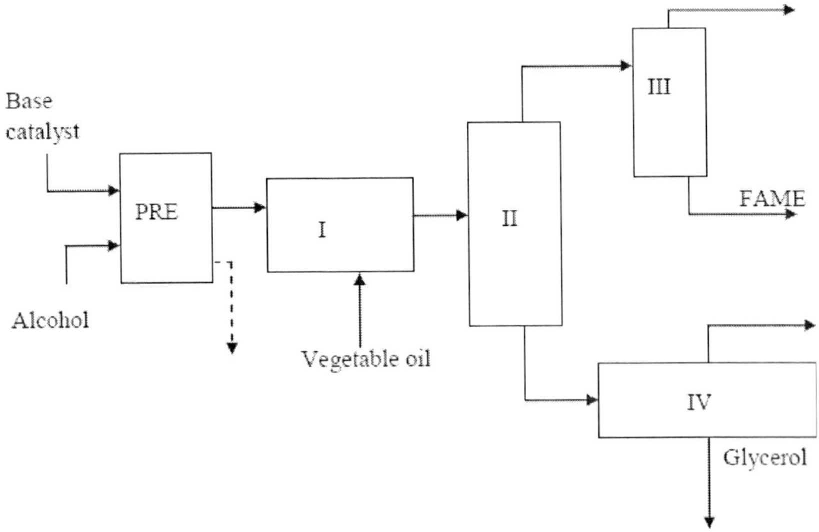

Figure 1. Typical process to produce biodiesel with a preesterification step.

After the reaction with the vegetable oil, (equipment I) separation of the glycerol phase and the biodiesel phase is required (equipment II). Several purifications are needed for both streams: washing with water, neutralization of the base catalyst, distillation of the raw materials that have not reacted, etc are perform in order to recover raw materials. To separate waste and effluents from the main product but also to produce FAME and glycerol according to international standards in order to sell them (Equipment III and IV summarize this purification step).

The reaction taking place in equipment I is the transesterification, in this case direct esterification will never take place since in the presence of the basic catalyst the free fatty acid will interact to produce soaps, according to the following reaction [5,8].

$$R'\text{-}COOH + Na(OH) \leftrightarrow R'\text{-}COO\text{-}Na + H_2O \qquad (2)$$

This reaction is faster and has several inconveniences to the biodiesel process:

- It could consumes a possible raw material, the free fatty acid
- It consumes the catalyst, leaving a lower amount available to produce the alkoxy that will allow the transesterification to take place.

- It will produce soap that is a non desirable product. Although it might have a selling market, the purification and separation from the glycerol and biodiesel is complicated and will increase the production costs.
- In reaction 2 water is produced, due to the presence of it, triglycerides might suffer hydrolysis, making more free fatty acid that will interact with more catalyst to generate more soap, this process consume raw material, catalyst, and created undesirable products.

Because of all this, the typically raw material for the basic homogeneous process is the refine oil, where the amount of impurities is completely reduced. The amount of free fatty acid is lower than 0.1 to 0.5 % [1-2,5,8], while humidity should be lower than 0.1%, the amount of phosphorus has a maximum in 20-50 ppm and the un-soaponificable material should be below 1%, among other restrictions imposed by the international standards.

2. Transesterification Reaction

As previously mentioned, this reaction is the interaction of a triglyceride and an alcohol in the presence of a catalyst, in this case, a basic catalyst. The reaction has three steps, and how the catalyst interacts in each of them is summarizes as follows [4,9]:

From triglycerides to diglycerides (Equation 3):

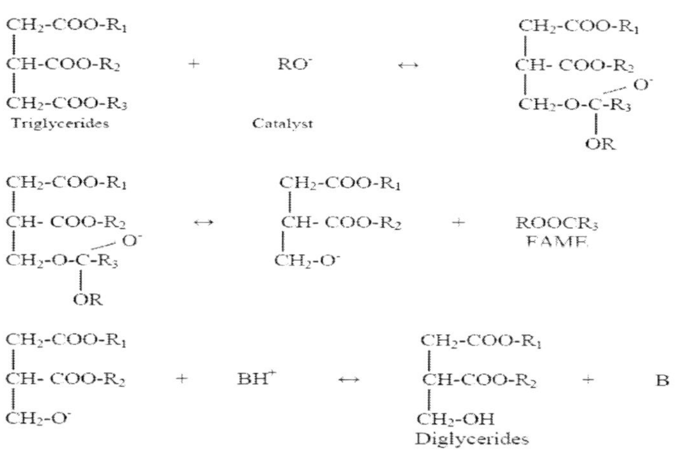

From diglycerides to monoglycerides (Equation 4)

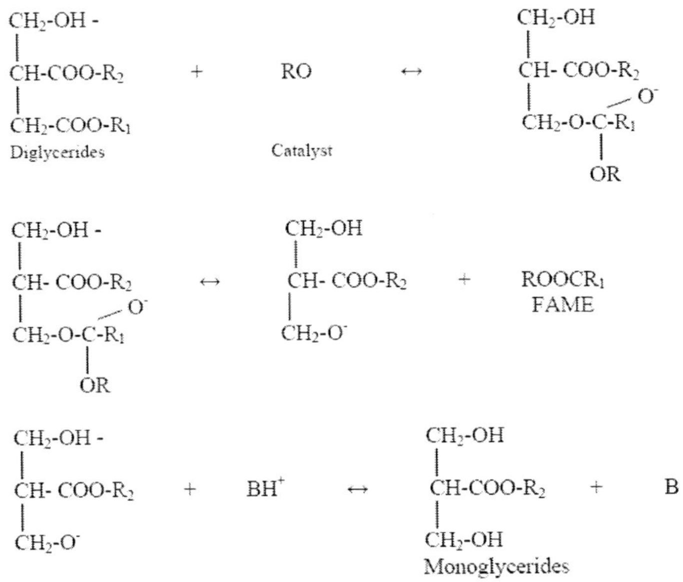

And from monoglycerides to glycerol, (Equation 5)

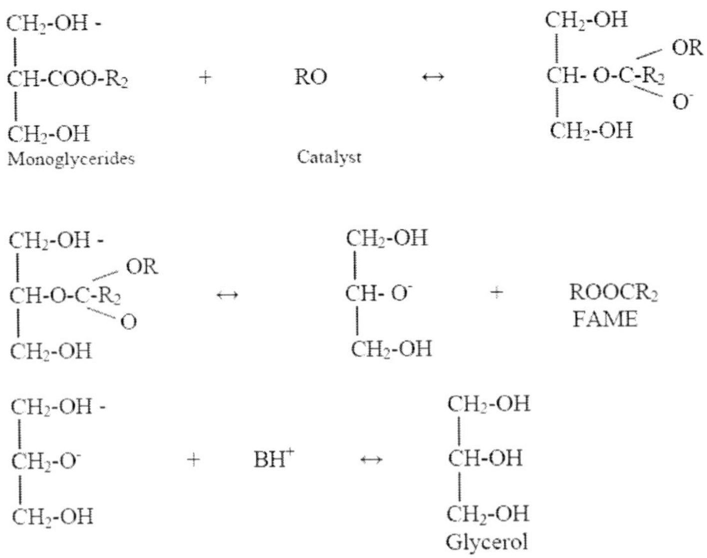

As it will be point out later on this chapter, several operational conditions could be found and have been established to be accurate to carry on the transesterification reaction. It is not the purpose of this book to give less credit to any of them but of contrarily, it will be tried to focus on the results from each of them and compared with each other their differences and strength.

Several authors have worked on this reaction over a long time so far and a lot of possible scenarios have been described. Tomasevic and Marinkovic [10] have carried out the transesterification reaction using potassium hydroxide as catalyst and methanol and frying oil as raw material. They have found that for low temperature (25°C), a standard molar ratio alcohol/oil = 6:1 and for 30 minutes reactions the biodiesel yield was big enough to provided a biodiesel to be used in regular engines. Noureddini and Zhu [11] have done a similar work, they have studied the transesterification reaction in the presence of methanol but also focusing in the Reynolds number related to the mixing intensity. The operational conditions were slightly different from the previous one; a reaction temperature of 70°C was used with a catalyst amount of 0.2% wt. However, the molar ratio was of 6:1. In that work two agitations Reynolds number were tested with good results. A kinetic model was also obtained and the experimental information was well predicted.

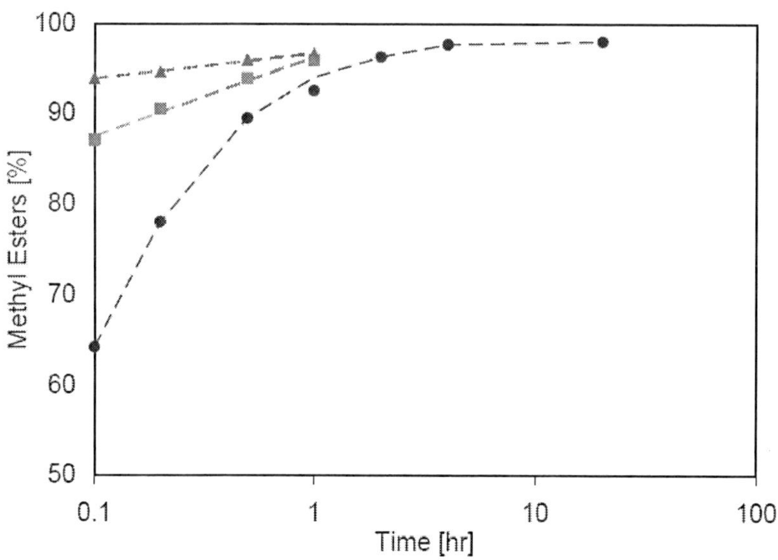

Figure 2. Variations of the biodiesel amount due to changes in the reaction temperature as a function of time. Molar ratio of 6:1, 1% wt. of catalyst. (●) 32°C, (■)45°C and (▲) 60°C.

Freedman et al. [12], carried out the same reaction but with a lower temperature (32°C), but also with a higher one (60°C). They have found that under similar molar ratio (6:1), and with the same amount of catalyst load, the difference in the reaction temperature it is seen on the time required to achieved full conversion, requiring 1 hr for 60°C and around 4hr for the lower temperature. A comparison of these results could be seen on Figure 2, extracted from reference [12].

Several potassium and sodium based homogeneous catalyst have been tested by Vicente et al. [13], showing that methoxide produces a higher final biodiesel yield under similar operational conditions. In all cases the purity of the biofuel is close to 100% after 4 hours of reaction. The reaction temperature was close to the previously mentions (65°C) and with a molar ratio of 6:1. In their work the mention how the sample was extract from the reaction medium and prepared for analysis. Figure 3, extracted from reference [13] shows the behavior of this reaction for the four tested catalyst. It can be seen the full conversion achieved for sodium hydroxide in less than 40 minutes.

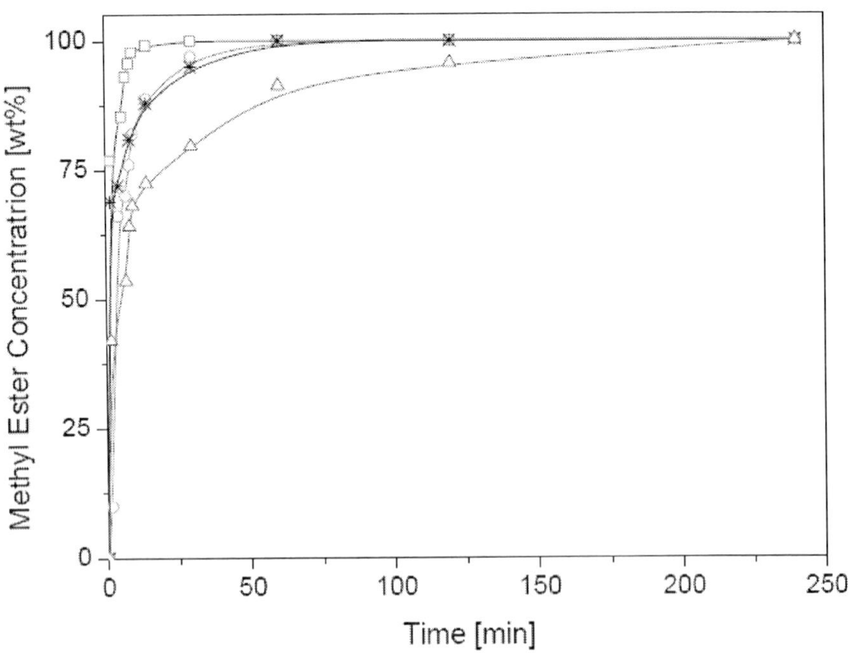

Figure 3. Variation of the percentage of biodiesel produced for several catalysts over time. (□) Sodium hydroxide, (○) Potassium hydroxide, (*) Sodium methoxide, (Δ) Potassium methoxide.

Meng et al. [14] have performed the transesterification reaction with methanol using waste cooking oil in the presence of alkaline catalyst. They have used a higher molar ratio (9:1), and a slightly smaller temperature (50°C), they have reached a conversion of 89.8% in 90 minutes. No comments on the amount of the possible un-reactant material. However, it was proved that this biodiesel could be blend with regular diesel and generated a useful blend.

Many other works might be found in the literature regarding this catalyst, to name some, it is possible to find the work done by Georgogianni et al. [15], Alamu et al. [16], Jeong et al. [17], Dias et al [18], and many other, where the reaction of different vegetable oils (triglycerides) and methanol was carried on in the presence of homogeneous basic catalyst.

Despite that the mayor used alcohol is methanol due to its lower price; ethanol is also a possible candidate to produce biodiesel. The use of this alcohol it has some advantages over using methanol being two of the most important its low toxicity for humans, that allow a simpler handling and storage but also that it might be produce from renewable source, making a final biodiesel closer to be 100% natural, renewable and sustainable.

Some works related to the use of ethanol are: the work done by Alamu et al. [19], they have studied the influence of several molar ratios in the transesterification reaction of palm kernel oil. The reaction was done at the regular operational conditions of T = 60°C, 1% wt. of catalyst (it was used potassium hydroxide) and the reaction took place for 2 hours. They have tested several concentrations of alcohol and the results could be seen in Table 1, obtained from reference [19].

Table 1. Variations of the biodiesel yield as a function of the molar ratio

Initial molar ratio	Final yield [%]
0.100	29.5
0.125	54.0
0.150	75.0
0.175	89.0
0.200	96.0
0.225	93.5
0.250	87.2

Table 2. Final yield of biodiesel for several operational conditions

Temperature [°C]	Molar ratio	Catalyst load [% wt.]	Catalyst	Yield [%]
35	12:1	1.0	KOH	61.1
60	12:1	1.0	KOH	72.5
78	12:1	1.0	KOH	74.2
60	12:1	1.0	CH3ONa	≈60
60	12:1	1.0	CH3OK	≈65
60	12:1	1.0	NaOH	≈64
60	12:1	0.5	KOH	≈55
60	12:1	1.5	KOH	≈70
60	6:1	1.0	KOH	≈45
60	9:1	1.0	KOH	≈60
60	15:1	1.0	KOH	66.2

Zhou et al. [20] have, as well, analyzed the transesterification reaction using ethanol and different vegetable oil. In their work the authors compared a typical production way with a new alternative using tetrahydrofuran as co-solvent with the aim of reaching a homogeneous system allowing the glycerol to stay in the same phase as the biodiesel. They have found that in the presence of a co-solvent, a molar ratio of 25:1 is needed in order to reach good final results. Several operational conditions were tested. The higher yield achieved was 99.4% at 23°C, 25:1 of molar ratio (methanol/oil), 1.4 % wt. of catalyst KOH and the required amount of co solvent to produce a miscible system with the alcohol.

Similar research was done by Kucek et al. [21] where the transesterification of refine oil was done in the presence of sodium and potassium hydroxide. They have reached a conversion of 97.2% in similar operational conditions as the previous work, but with a lower temperature range and different molar ratios. However, as with the previous research, ethanol was used as alcohol. For this study the authors have used an experimental design study, allowing not only the understanding of the reaction, but also to have a higher interaction among the studied variables. Testing similar catalyst as those by Kucek et al. [21], Encinar et al. [22] have studied the transesterification reaction using ethanol and sodium and potassium hydroxide and their methoxide as catalyst (as done by Vicente et al. [13]). Reaction temperature, molar ratio, catalyst load, type of catalyst were the studied variables. A summary of the result presented in their work could be seen in Table 2, where all the previous mentioned operational conditions are showed as well as the final yield achieved for each scenario. From Table 2 it can be seen that the mayor conversion is obtained for T = 78°C , Molar ratio = 6:1 , and a catalyst load of 1%.

From all the previous work, one might be able to establish that when methanol is used and sodium hydroxide is employ as catalyst, a typical operational condition to be use are: reaction temperature of 60°C, 1%wt. of catalyst and a molar ratio of 6:1, giving a full conversion in around 2 hours. Rashid et al. [23] have done an optimization studied of those major variables for a sunflower methyl ester process. Several temperatures as well as molar ratios and catalyst load have been tested obtaining as results that the higher conversion was achieved under the above mentioned operational conditions. Table 3 shows that the minimum yield achieved was 51.6% that correspond for the smallest amount of catalyst, meaning that might be required a higher reaction time to achieve a higher final conversion.

Table 3. Sunflower transesterification reaction for several conditions with a reaction time of 2 hr, extracted from [23]

Reaction temperature [°C]	Molar ratio methanol/oil	Catalyst load [wt%]	Catalyst type	Yield [%]
60	6:1	1.00	KOH	86.7
60	6:1	1.00	$KOCH_3$	90.0
60	6:1	1.00	$NaOCH_3$	82.7
30	6:1	1.00	NaOH	90.9
45	6:1	1.00	NaOH	92.8
60	6:1	1.00	NaOH	97.1
60	3:1	1.00	NaOH	61.5
60	9:1	1.00	NaOH	93.0
60	12:1	1.00	NaOH	86.0
60	15:1	1.00	NaOH	83.8
60	18:1	1.00	NaOH	81.0
60	6:1	0.25	NaOH	51.6
60	6:1	0.50	NaOH	65.0
60	6:1	0.75	NaOH	92.9
60	6:1	1.25	NaOH	82.7
60	6:1	1.50	NaOH	78.1

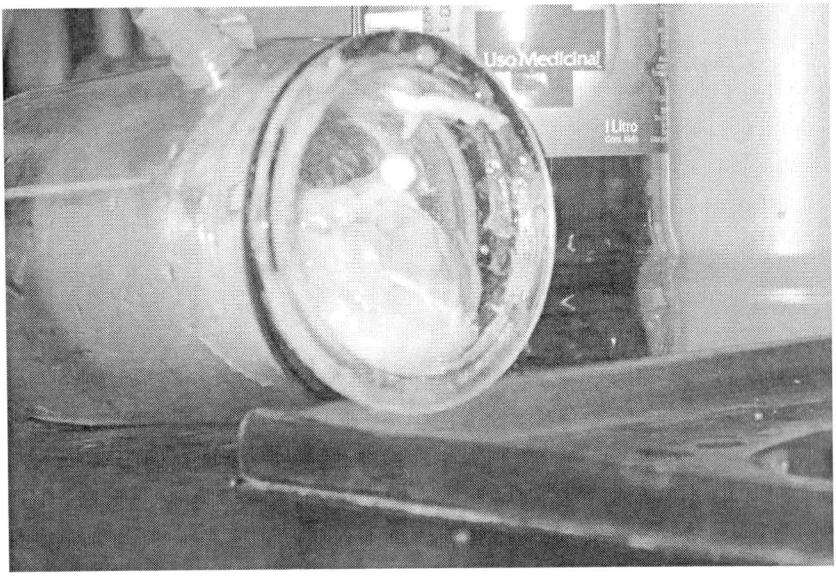

Figure 4. Semi solid mixture of acid oil and sodium hydroxide.

Besides the production of biodiesel using conventional mixing, stirring reactors, etc. Some work addresses the possibility of using new technologies to carry on the process. Hanh et al. [24] performed a methanolysis reaction, under ambient conditions but using ultrasonic irradiation. The operational conditions were standards ones (T=60°C), molar ratio of 6:1, 1 % wt. of catalyst. With regular mixing the final conversion was achieved in 4 hr, while with ultrasonic irradiation they achieved similar results after 30 minutes of reaction. A similar technique was used by Colucci et al. [25], they have used ultrasonic mixing to transformed soybean oil into biodiesel in less than 15 minutes at low temperature and regular molar ratio of methanol/oil. The final conversion achieved was 99.4%.

Azcan and Danisman [26-27] have studied the production of biodiesel using microwave irradiation, they have found that a good quality biodiesel could be produced using this technique. The reduction in the reaction time is one of the most important results achieved, since the reaction was complete after 7 minutes. Under similar operational conditions but using conventional heating, the reaction will take around 30 minutes. One option to improve separation of the reactants and the product was established by Dubé et al. [28], who have used a membrane reactor in order to separate the fuel produced from the rest in order to force the reaction towards the desirable product.

Generally, one reactor might be enough to achieve the final conversion; nevertheless, in order to improve this, more than one reactor it is commonly used, generally it is two or three reactors [29-30].

As mentioned at the beginning of the chapter, the basic homogeneous catalyst process is suitable and gives good results when a refine oil is used, or if some purification is done before the transesterification reaction takes place. However, in order to evaluate the reaction using a lower quality feedstock, some experiments were done with potassium hydroxide and sodium hydroxide but with an acid oil (initial amount of FFA = 10% wt.). It was found that when sodium hydroxide was employed, a semi solid mixture was produce (see Figure 4) most likely to be soap; however, the reaction had to be stopped since stirring was impossible to continue.

Contrarily, when potassium hydroxide was used, soap were definitively produced, since the conditions are good for this reaction to happen, but it seems that they are soluble in the reaction medium, allowing the reaction to continue. The transesterification reaction was definitively slower due to a important loose of catalyst but was possible to continue. After the reaction was finished, the mixture was decanted and centrifuge and it was seen that more than half of the reactor was fulfill with soap or a mix of soaps, as it could be seen in Figure 5.

Figure 5. Mixture of oil and soap after reaction when using potassium hydroxide.

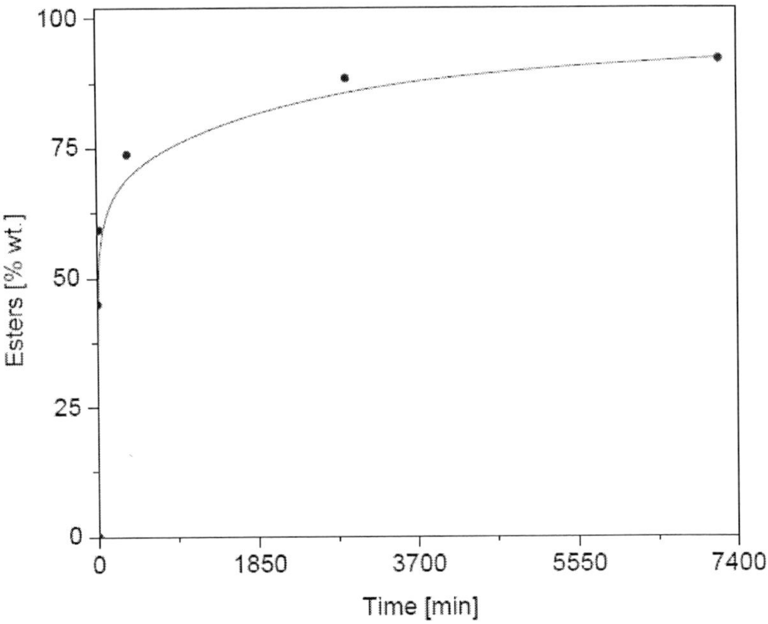

Figure 6. Variation of the % of esters produced when using an acid oil (10% of FFA) and potassium hydroxide as catalyst in the presence of ethanol.

From the upper phase, samples were taken and analyzed by gas chromatography in order to see the production of esters. Figure 5 shows the variation of the ester production for KOH as catalyst under similar operational conditions as previously mentioned.

It is important to see that a conventional transesterification reaction of a refine oil will give over 98% yield in 2 hr. For this case, we reached the top conversion of 92% in over 120 hr (5 days). It could be seen from Figure 6 that even thought the final conversion is high; it is not high enough to satisfy international standards. Even more, if the desirable final conversion would have been reached, the operational reaction time would have been too big to be used in an industrial scale.

All the previous works were carried on generally using vegetable oil, however, animal fat is another possibility to produce biodiesel. However, this feedstock is still under development due to difference on their properties compared to vegetable oil. Some work using this type of feedstock could be found, for example, Chung et al. [31] and Özner and Altun [32], among other, showing that it is a suitable, raw material for biodiesel production.

3. KINETICS

Despite the technological viability for producing biodiesel, as described before, it is important to study and understand the kinetic of the reaction in order to be able to predict the behavior under different operational conditions. This type of study, however, has the drawback that are attach to a system. Even though it is possible to extrapolate results to higher concentration of reactants, or different temperature or amount of catalyst, it is not possible to change any of them without introducing an error to our simulation; for example, when using ethanol for the experiments, the reaction time it is different than the one obtained when using methanol, if the kinetic is obtain for methanol, the use of ethanol will introduce a small error. However, general tendencies might be similar.

Because of this, several possible kinetic expression and kinetic parameter could be found in the open literature. Here I will show some of them for different scenarios and how they reproduce the experimental information. Bambase et al. [33], Noureddini and Zhu [34] and Darnoko and Cheryan [35] have done a similar study of the kinetics of the methanolysis reaction of different vegetable oils using KOH as catalyst. Bambase et al. [33] have used crude sunflower oil and the results obtained are in the acceptable range compared to those obtained by Darnoko and Cheryan [35] as well as Noureddini and Zhu [34], Table 4 shows the comparison of the activation energy for this three works, Table 4 was extracted from reference [32].

Despite the fact that the kinetic parameters were similar, different kinetic models were proposed for each case due to the difference in the initial system. Darnoko and Cheryan [35] have used a pseudo second order model for initial times while the reaction follows a zero order kinetics. On the other hand the works done by Bambase et al. [33] and Noureddini and Zhu [34] follows a similar development, both considered the reaction showed in equation 1 to 3 to be elementary step. Noureddini and Zhu [34] have also included the overall reaction for the determination of the kinetic parameters while Bambase et al. [33] have considered only the reactions in equation 1 to 3.

The authors were able to reproduce experimental information for changes in temperature, amount of catalyst, molar ratio and mixing intensity, showing in all cases a good agreement between the experimental information and the kinetic model.

Table 4. Comparison of the activation energy

Reaction	Activation Energy		
	Bambase et al. [33]	Noureddini and Zhu [34]	Darnoko and Cheryan [35]
TG→DG	14040	13145	14700
DG→TG	10739	9932	--------
DG→MG	16049	19860	14200
MG→DG	13907	14639	--------
MG→GL	7173	6421	6400
GL→MG	10997	9588	--------

The model obtained by Bambase et al. [34] it is reproduce here below:

$$
\begin{aligned}
\frac{dC_{tg}}{dt} &= -k_1 C_{tg} C_{al} + k_2 C_{dg} C_{me} \\
\frac{dC_{dg}}{dt} &= k_1 C_{tg} C_{al} - k_2 C_{dg} C_{me} + k_4 C_{mg} C_{me} - k_3 C_{dg} C_{al} \\
\frac{dC_{mg}}{dt} &= k_3 C_{dg} C_{al} + k_6 C_{gl} C_{me} - k_4 C_{mg} C_{me} - k_5 C_{mg} C_{al} \\
\frac{dC_{me}}{dt} &= k_1 C_{tg} C_{al} + k_3 C_{dg} C_{al} + k_5 C_{mg} C_{al} - k_6 C_{gl} C_{me} - k_4 C_{mg} C_{me} - k_2 C_{dg} C_{me} \\
\frac{dC_{al}}{dt} &= -\frac{dC_{me}}{dt} \\
\frac{dC_{gl}}{dt} &= -k_6 C_{gl} C_{me} + k_5 C_{mg} C_{al}
\end{aligned} \qquad (6)
$$

A similar work, based on Noureddini and Zhu model [34] was done by Narváez et al. [36] where the shunt reaction was used and the kinetics parameter for that reaction were also calculated. From their results it was found that the reaction rate constant was either zero or of the order of 10^{-6} for the shunt reaction, while for the rest they were of the order of 10^{-1}, showing that the shunt reaction has almost no influence on the system, converging to the model proposed by Bambase et al. [33]. Nevertheless, the experimental information showed by Narváez et al. [36] it is well reproduce by the kinetic model showed by them or by Noureddini and Zhou [34].

Based on the same reaction as Bambase et al. [33], without the presence of the shunt reaction, Vicente et al. [37] obtained a kinetic mechanism for the transesterification reaction considering that the conversion of the raw material into product could take place due to a catalyzed reaction as well as a non-catalyzed reaction. This combination of effects was considered in the reaction rate constant, defined as follows:

$$k_i^{'} = k_i + Ck_{i0} \tag{4}$$

Using this correction to the rate constant, the authors achieved a good reproducibility of their results for all the range studied as published in reference [37].

Cao et al. [38] have developed a similar kinetic expression for the homogeneous transesterification reaction of canola oil but using sodium hydroxide as catalyst and with a membrane reactor instead of a stirring system. The membrane reactor produces a continuous separation of the product, making that the final expression will be modified by this physical separation of compounds. The kinetic model is based on expressions 1 to 3 considering them as elementary steps and in their model the authors have also introduced a modification established by Vicente et al. [37] previously described. The final expression obtained by Cao et al. [38] introduces the flow of some of the products throw the membrane and therefore there is an accumulation or some of the products in each size of the membrane.

The authors have obtained a kinetic constant for the reaction taking place at 65°C and have compared with the work of Vicente at al. [37,39], it is establish that the reaction constant are similar except for those involving the conversion of diglycerides to monoglycerides and the backwards reaction. The higher value on the constant rate (22.46 [l/(mol·min)] for Cao et al. [38] to 2.06 [l/(mol·min)] for Vicente et al. [37] and 0.77 [l/(mol·min)] for Vicente et al.[39]) it might be to a higher molar ratio of alcohol in the membrane reactor (24:1) compared to those used in a steering reactor (6:1). Nevertheless, the comparison between the experimental data and the model proposed by Cao et al. [38] shows good agreement.

In the work done by Komers et al. [40], a kinetic mechanism is based on two possible reactions competing to each other, the methanolysis of triglycerides and the saponification of the compounds by the catalyst. For the transesterification reaction the author have not used the elementary expression as previously done, it was assumed a simple mechanism where the transformation of triglycerides into diglycerides is divide in two steps, the same for diglycerides and monoglycerides. Using this mechanism, the experimental information was good correlated.

4. Economic Study

Economic feasibility is very relevant in order to see if the technological research could be scale up to industrial production. Several plants are working nowadays producing biodiesel, and most of them are using basic homogeneous catalyst such as sodium hydroxide or sodium methoxide. Therefore, its economical viability is good; however, it is important to distinguish if there is any kind of economic motivation by the government in order to produce biofuel, such as taxation reduction, or financial aid, or even national law imposing the use of some biodiesel blended with regular diesel. All this factor will also produce a difference in the economic scenario of a biodiesel plant and will be important factor at the moment of the economic analysis.

A very complete study of an homogeneous basic plant was done by Haas et al. [30], where several variables and a good description of the process is done. The authors have considered several parts of the plant and in good detail all the cost and prices involved. Other's analysis for different technologies will be introduced in the following chapter as each alternative is described from a technical point of view.

References

[1] Ma, F. Hanna, M. A. *Biodiesel production: a review. Bioresource Technology*. 70. (1999), 1-15.

[2] Fukuda, H. Kondo, A. Noda, H. Biodiesel *Fuel Production by Transesterification of Oils*. J. *Bioscience and Bioengineering*. 92(5). (2001), 405-416.

[3] Barnwal, B.K. Sharma, M.P. *Prospects of Biodiesel production from vegetable oils in India. Renewable and Sustainable Energy Reviews*. 9(4). (2005), 363-378

[4] Srivastava, A. Prasad, R. *Triglycerides-based diesel fuels. Renewable and Sustainable Energy Reviews*. 4. (2000), 111-133.

[5] Marchetti, J.M., Miguel, V.U., Errazu, A.F. *Possible methods for biodiesel production. Renewable and Sustainable Energy Reviews*. 11. (2007), 1300-1311.

[6] Schuchardt, U., Sercheli, R., Vargas, R.M. *Transesterification of vegetable oils: a review. JBCS*, 9(1). (1998), 199–210.

[7] Knothe, G., Van Gerpen, J., Krahl, J. (Eds.), *The biodiesel handbook*, AOCS Press, Champaign, Illinois, 2005, 302 pp.

[8] Marchetti, J.M. Ph.D. thesis. *Technological alternatives for biodiesel production*. Universidad Nacional del Sur. Argentina (2008).

[9] Drapcho, C.M., Nhuan, N.Pm Walker, T.H. *Biofuels Engineering Process Technology*. McGraw Hill, USA, 2008.

[10] Tomasevic, A.V., Siler-Marinkovic, S.S. *Methanolysis of used frying oil. Fuel Processing Technology*. 81. (2003), 1-16.

[11] Noureddini, H., Zhu, D. *Kinetics of transesterification of soybean oil. Journal of American Oil Chemists Society*. 74(11). (1997), 1457-1463.

[12] Freedman, B., Pryde, E.H., Mounts, T.L. *Variables affecting the yields of fatty esters from transesterified vegetable oils. Journal of American Oil Chemists Society*. 61(10). (1984), 1638-1643.

[13] Vicente, G., Martínes, M., Aracil, J. *Integrated biodiesel production: a comparison of different homogeneous catalysts systems. Bioresource Technology*. 92. (2004), 297-305.

[14] Meng, X., Chen, G., Wang, Y. *Biodiesel production from waste cooking oil via alkali catalyst and its engine test. Fuel Processing Technology*. 89. (2008), 851-857.

[15] Georgogianni, K.G., Katsoulidis, A.K., Pomonis, P.J., Manos, G., Kontominas, M.G. *Transesterification of rapeseed oil for the production of biodiesel using homogeneous and heterogeneous catalysis. Fuel Processing Technology*. DOI: 10.106/j.fuproc.1009.03.002. (article in press)

[16] Alamu, O.J., Waheed, M.A., Jekayinfa, S.O. *Biodiesel production from Nigenerian palm kernel oil: effect of KOH concentration on yield. Energy for Sustainable Development*. XI(3). (2007), 77-82.

[17] Jeong, G.T., Park, D.H. *Batch (one- and two-stage) production of biodiesel fuel from rapeseed oil. Applied Biochemistry and Biotechnology*. 129-132. (2006), 668-679.

[18] Dias, J.M., Alvim-Ferraz, M.C.M, Almeida, M.F. *Comparison of the performance of different homogeneous alkali catalysts during transesterification of waste and virgin oils and evaluation of biodiesel quality. Fuel*. 87. (2008), 3572-3578.

[19] Alamu, O.J., Waheed, M.A., Jekayinfa, S.O. *Effect of ethanol-palm kernel oil ratio on alkali-catalyzed biodiesel yield. Fuel.* 87. (2008), 1529-1533.

[20] Zhou, W., Konar, S.K., Boocock, D.G.B. *Ethyl Esters from the single-phase base catalyzed ethanolysis of vegetable oils. Journal of the American Oil Chemists Society.* 80(4). (2003), 367-371.

[21] Kucek, K.T., César-Oliveira, M.A.F., Wilhelm, H.M., Ramos, L.P. *Ethanolysis of refined soybean oil assisted by sodium and potassium hydroxides. Journal of American Oil Chemists Society.* 84. (2007), 385-392.

[22] Encinar, J.M., González, J.F., Rodríguez-Reinares, A. *Ethanolysis of used frying oil. Biodiesel preparation and characterization. Fuel Processing Technology.* 88. (2007), 513-522.

[23] Rashid, U., Anwar, F., Moser, B.R., Ashraj, S. *Production of sunflower oil methyl esters by optimized alkali-catalyzed methanolysis. Biomass and Bioenergy.* 32. (2008), 1202-1205.

[24] Hanh, H.D., Dong, N.T., Starvarache, C., Okitsu, K., Maeda, Y., Nishimura, R. *Methanolysis of triolein by low frequency ultrasonic irradiation. Energy Conversion and Management.* 49. (2008), 276-280.

[25] Colucci, J.A., Borrero, E.E., Alape, F. *Biodiesel from an alkaline transesterification reaction of soybean oil using ultrasoni mixing. Journal of the American Oil Chemists Society.* 82(7). (2005), 525-530.

[26] Azcan, N., Danisman, A. *Alkali catalyzed transesterification of cottonseed oil by microwave irradiation.* Fuel. 86. (2007), 2639-2644.

[27] Azcanm N., Danisman, A. *Microwave assisted transesterification of rapeseed oil. Fuel.* 87. (2008), 1781-1788.

[28] Dubé, M.A., Tremblay, A.Y., Liu, J. *Biodiesel production using a membrane reactor. Bioresource Technology.* 98. (2007), 639–647.

[29] Assman, G., Blasey, G., Gutsche, B., Jeromin, L., Rigal, J., Armengand, R., Cormary, B. *Continuous progress for the production of lower alkayl esters.* US patent 5514820 (1996).

[30] Haas, M.J., McAloon, A.J., Yee, W.C., Foglia, T.A. *A process model to estimate biodiesel production costs. Bioresource Technology.* 97(4). (2006), 671-678.

[31] Chung, K.H., Kim, J., Lee, K.Y. *Biodiesel production by transesterification of duck tallow with methanol on alkali catalysts. Biomass and Bioenergy.* 33. (2009), 155-158.

[32] Özner, C., Altum, Ş. *Biodiesel production from inedible animal tallow and an experimental investigation of its use as alternative fuel in a direct injection diesel engine. Applied Energy*. 86. (2009), 2114-2120.

[33] Bambase Jr., M.E., Nakamura, N., Tanaka, J., Matsumura, M. *Kinetics of hydroxide-catalyzed methanolysis of crude sunflower oil for the production of fuel-grade methyl ester. Journal of Chemical Technology and Biotechnology*. 82. (2007), 273-280.

[34] Noureddini, H., Zhu, D. *Kinetics of transesterification of soybean oil. Journal of the American Oil Chemists Society*. 74(11). (1997), 1457-1463.

[35] Darnoko, D., Cheryan, M. *Kinetics of palm oil transesterification in a batch reactor. Journal of the American Oil Chemists Society*. 77(12). (2000), 1263-1267.

[36] Narváez, P.C., Rincón, S.M. Sánchez, F.J. *Kinetics of palm oil methanolysis. Journal of the American Oil Chemists Society*. 84. (2007), 971-977.

[37] Vicente, G., Martínez, M., Aracil, J., Esteban A. *Kinetics of sunflower oil methanolysis. Industrial and Engineering Chemistry Research*. 44. (2005), 5447-5454.

[38] Cao, P., Tremblay, A.Y., Dubé, M.A. *Kinetics of canola oil transesterification in a membrane reactor*. Industrial and Engineering Chemistry Research. 48. (2009), 2533-2541.

[39] Vicente, G., Martínez, M, Aracil, J. *Kinetics of Brassica carinata oil methanolysis. Energy and Fuel*. 20. (2006), 1722-1726.

[40] Komers, K., Skopal, F., Stloukal, R., Machek, J. *Kinetics and mechanism of the KOH-catalyzed methanolysis of rapeseed oil for biodiesel production. European Journal of Lipid Science and Technology*. 104. (2002), 728-737.

Chapter 5

HOMOGENEOUS ACID TECHNOLOGY

1. INTRODUCTION

As mentioned in Chapter 3, the most common raw material used for biodiesel production is refine oil. The mayor disadvantage is its high price [1-2] due to the refining process. One solution to reduce the cost of the raw material in a biodiesel plant is to use lower quality raw materials, such as acid, frying, waste or cooking oil. However, in each of the previous option, the amount of impurities is much higher than the acceptable for the conventional biodiesel production process [3-4].

Because of that, several options have been developed. One of them is the use of an acid catalyst such as sulfuric acid [5-8]. This catalyst could be used in two different ways:

1. As the regular catalyst to carry on the transesterification reaction.
2. As a pretreatment option before a basic homogeneous catalyst production process.

Each case has the advantages and drawbacks regarding reaction time, temperature, molar ratios, etc.

Even thought it will be possible to use cheaper raw materials, the down streaming separation and purification will continue to be complicated. Several equipments are required for washing the biodiesel and extract the impurities as well as neutralizing the catalyst and removing the un-reactant methanol. Glycerol needs to be separated from the biodiesel, obtaining it with a purity of industrial grade.

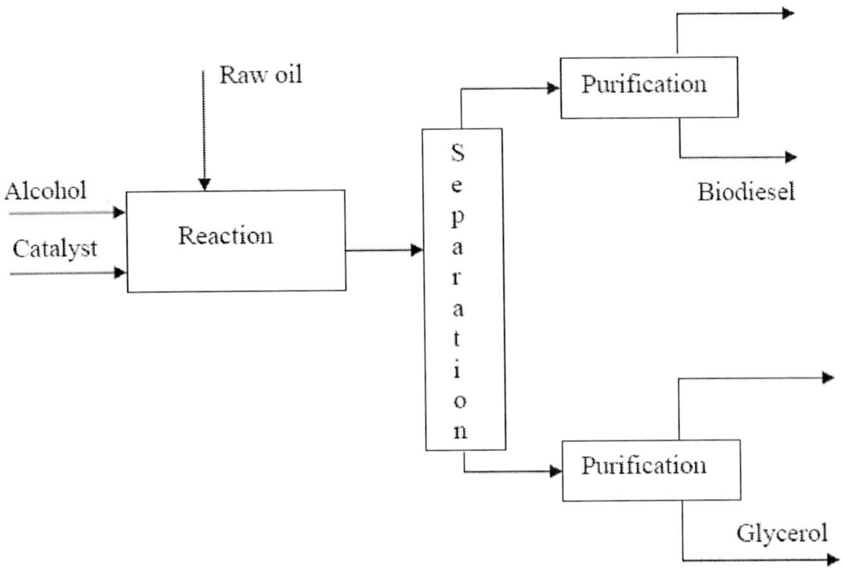

Figure 1. Schematic flow diagram of an acid catalyzed biodiesel process.

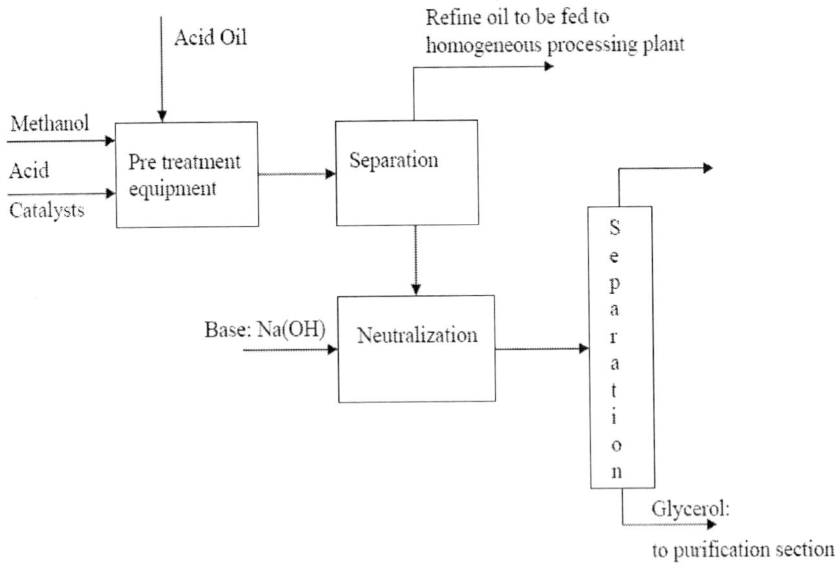

Figure 2. Pretreatment for the free fatty acids of an acid oil using an acid.

A schematic flow sheet of the process it is showed in Figure 1.

In Figure 1 it can be seen a process with a reaction section, a purification and separation. If acid catalyst is used, similar when using a base catalyst, purification after the reaction step is required: *i)* washing with water, *ii)* neutralizing catalyst, and *iii)* distillation separation are the mayor equipment and process involved in order to purified the biodiesel, to obtain a glycerol in the industrial grade, to make methanol to be fed back in the main equipment and to separate the effluents for their treatment [9-11].

This catalyst could be used as a pretreatment step for acid oils; this means that it will be a preesterification process before a basic homogeneous catalyst plant treats the oil to produce biodiesel, so both reactions are relevant in order to produce biodiesel. Therefore, they will be discussed separately with the purpose of a better understanding of each reaction and its influence on the global process. A typical flow sheet of this pretreatment process could be seen in Figure 2.

It can be seen that now the acid catalyst is only used in the pretreatment section. After this, the catalyst will be neutralized and then a new load of basic catalyst will be added to the system to continue with the transesterification reaction.

Due to the fact that both reactions are relevant and of great interest, they will be addressed separately in order to be able to provide more information for each of them. At the end of the chapter it will be a kinetic study of both reactions.

2. Esterification Reaction

The esterification reaction takes places when sulfuric acid is employed. Either as a pretreatment of a conventional biodiesel plant or as a part of the main reaction stage. In both cases, the conversion of free fatty acids into biodiesel will take place in the presence of a vegetable oil. This reaction could be simplified as follows:

$$R\text{-}COOH + R^1\text{-}CH_2\text{-}OH \rightarrow R\text{-}CO\text{-}O\text{-}CH_2\text{-}R^1 + H_2O \qquad (1)$$

Several studies, have been done related to the esterification of fatty acids under several operational conditions; however, this was done with pure fatty acid. Several authors have worked with acetic acid as primary raw material, they have carried out the reaction under several temperatures but also for different types of alcohol [12-17], other types of fatty acid were also used, such as Acrylic acid [18], Levulinic acid [12], Oleic acid [15], Palmitic acid [16], etc. Other variables have also been studied in those works, such as reaction temperature, molar ratio, type and amount of catalyst (in some cases HCl as well as the same fatty acid could be used as catalyst), it was found that if the same fatty acid is used as catalyst, and with a temperature of 120 °C, a final conversion over 99% was reached in more than a day with a molar ratio of 20:1. On the other hand, if sulfuric acid was used, the same final conversion was achieved but in 10 hours with a much smaller molar ratio and a reaction temperature of 65°C [12]

Besides of the previous work, where fatty acid were used as raw material, some works have been done using and acid oil, in these cases, the free fatty acid concentration is lower and the presence of triglycerides might have an effect in the conversion of free fatty acid into biodiesel. Berrios et al. [19] have performed the esterification reaction of a mixture of fatty acid in the presence of sunflower oil. The amount of fatty acid was in the range as what it might be found in frying oil. Several temperatures, as well as amount of catalyst and agitation speed were tested with the aim of improving the final conversion. They have found that the best conditions where when a molar ratio of 60:1, a reaction temperature of 60°C and 5 % wt. of catalyst was used.

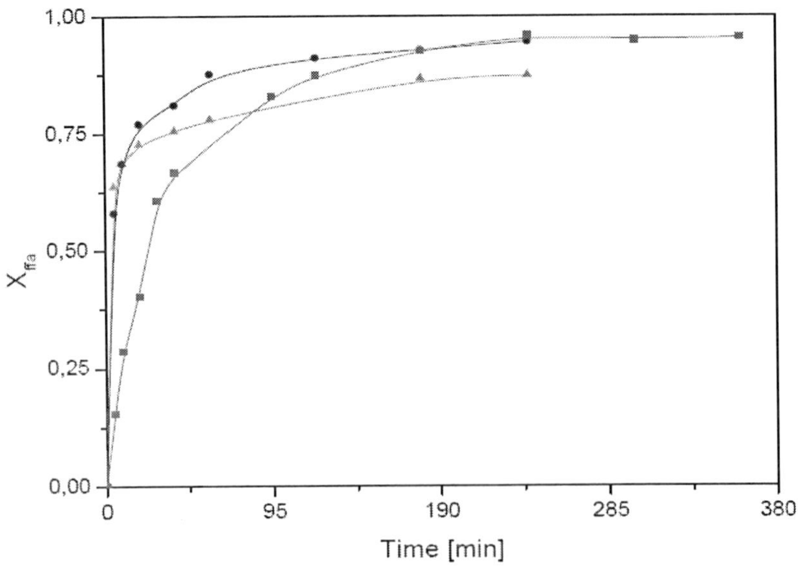

Figure 3. Variations of the conversion of free fatty acid as a function of the molar ratio. (▲) N = 4.21:1, (●) N = 6:12:1 and (■) N = 10.05:1.

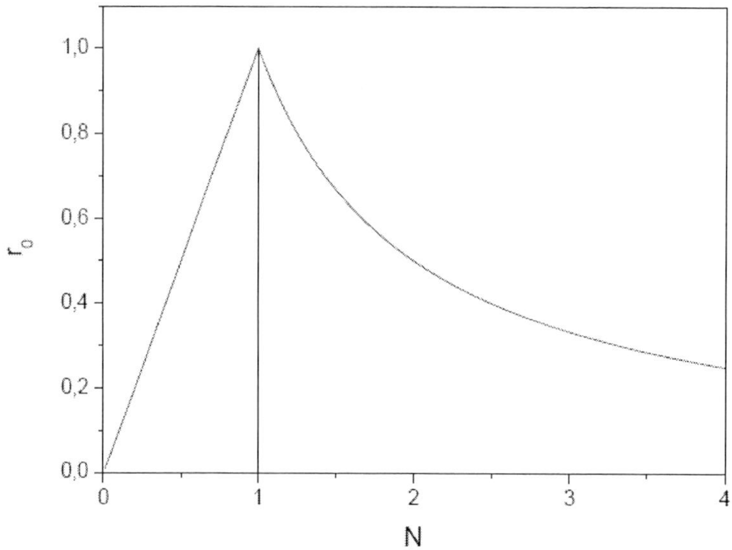

Figure 4. Variation of the initial reaction rate as a function of the molar ratio.

The authors have developed a kinetic model based on the reduction of the acidity number in the mixture. They have proposed a reversible reaction with some assumptions involved in order to correlate their data. The kinetics model fits considerable well. However, the influence of the triglycerides is not clearly stated.

Sendzikiene et al. [20] have studied the esterification reaction using oleic acid presented in a frying oil. Three different amounts of fatty acid were used varying from 0 up to 33% (w/w). They reported conversion that where around 50% for a 3 hr. reaction. Not mention of the influence of the triglycerides over the esterification reaction was found.

Marchetti and Errazu [21] have studied the esterification of oleic acid and ethanol in the presence of sunflower oil using sulfuric acid as catalyst. Several temperatures as well as molar ratio, amounts of catalyst and concentration of fatty acid were tested. It was found that at a 55°C reaction temperature, with a molar ratio of 6:1 and 2% wt. of sulfuric acid, a raw material that has 10%wt of oleic acid achieves a final conversion of 96% for the fatty acid in 4 h. Several comments are done in this work regarding the influence of the operational variables but not comments are explicit on the influence of the triglycerides due that a kinetic model is not presented there. Figure 3 shows the evolution of the conversion of free fatty acid when different amounts of alcohol are used in the system (extracted from reference [21]).

It can be seen that there is an unusual behavior of the conversion, being slower for higher concentration of alcohol and much faster for small concentrations of alcohol. It is believed that this could be due to two effects that are competing to each other, kinetics and a dissolution effect. When the concentration of alcohol is the biggest one, the dissolution is important and therefore the reaction is slower; however, since the final conversion is attach to the amount of the reactants, and then, this is the case where the higher final conversion is reached.

The behavior for the initial times could be easily approached if we considered a second order irreversible kinetic model:

$$r_0 = k \; C^0_{FFA} \; C^0_{ALC} \tag{2}$$

If the amount of alcohol is small, then the concentration of free fatty acid could be considered to be constant over time and expression 2 could be re writing in the following form:

$$r_0 = k^1 \; N \tag{3}$$

where $k^1 = k \; (C^0_{FFA})^2$ and $N = \dfrac{C^0_{ALC}}{C^0_{FFA}}$. From equation 3 it can be seen a straight tendency of the initial reaction rate with the molar ratio of alcohol to free fatty acid. On the other hand, if the alcohol is in great amount, then it might be considered constant and the initial reaction rate will take the following form.

$$r_0 = k^2 \; \frac{1}{N} \tag{4}$$

where $k^2 = k \; (C^0_{ALC})^2$ and $N = \dfrac{C^0_{ALC}}{C^0_{FFA}}$ .where the tendency is that of a hyperbolic function. If Equation 3 and 4 are plotted together (Figure 4), it can be seen that for small M, as the amount of alcohol increases the initial reaction rate increases as well; however, when a certain point is reached, increases on the amount of alcohol produces a decreasing in the initial reaction rate as seen also in the experimental data. The transition point is close to the stoichiometric molar ratio for the esterification reaction (N = 1)

3 Transesterification Reaction

As mentioned in the previous chapter, the transesterification reaction could take place when a base homogeneous catalyst is used. However, the oil has to be refined with small quantities of impurities. Another alternative catalyst is the acid homogenous one, generally used are sulfuric acid and toluene sulfonic. This alternative allows the treatment not only of refine oil but also crude, waste, frying, soapstocks and acid oil, where the quantity of impurities, free fatty acid for example, could be as high as possible, soapstocks could be over 80% of free fatty acid. However, there are two mayor disadvantages with this alternative:

1. The reaction time is 4 times higher than when NaOH is used, due to the fact that the reaction is around 4000 times slower.
2. The molar ratio of alcohol/oil is much higher, around 5 times higher than for NaOH, (normally a molar ratio of 32 is used [22-25])

The reaction mechanism is as follows [26-27]:

Triglycerides to diglycerides

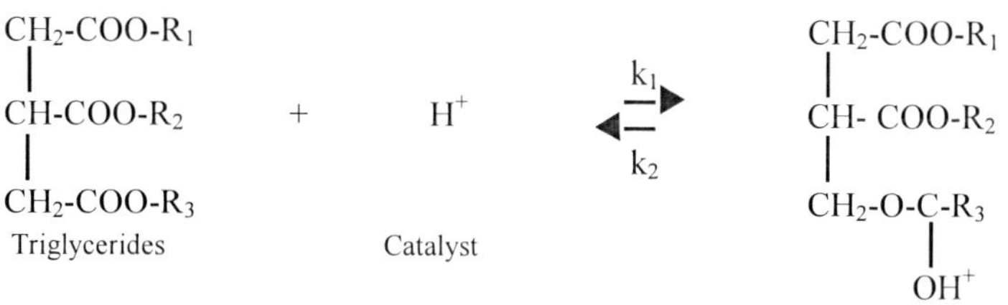

(5)

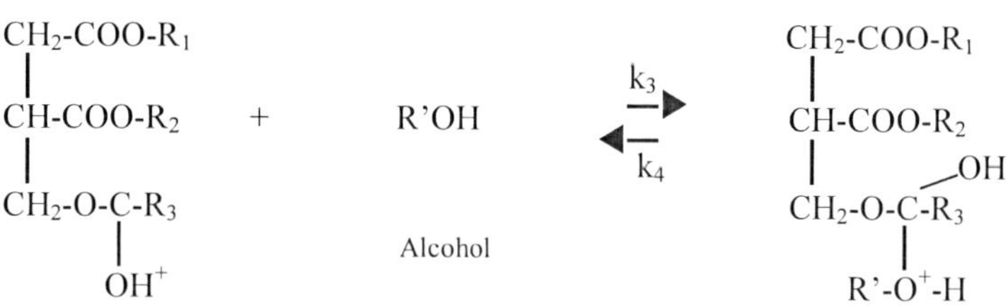

(6)

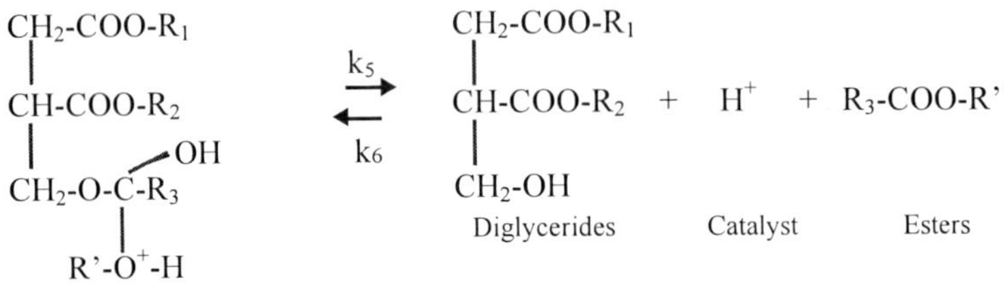

(7)

Diglycerides to monoglycerides

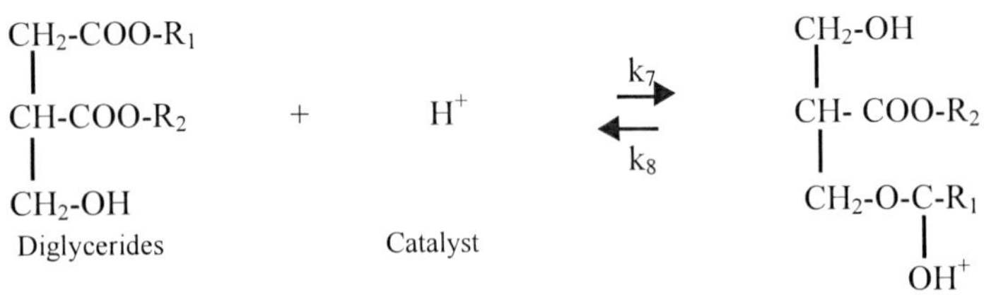

(8)

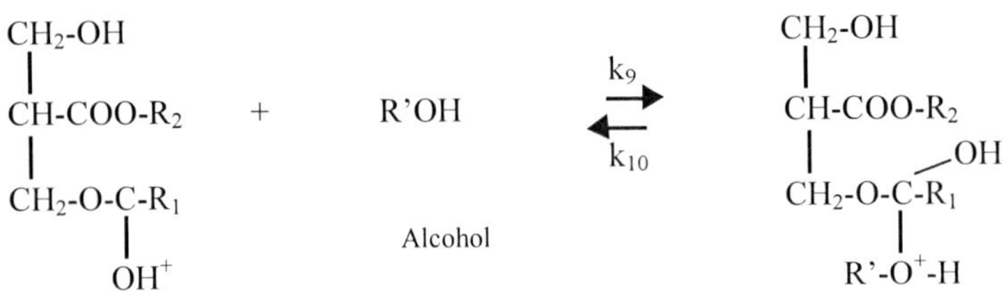

(9)

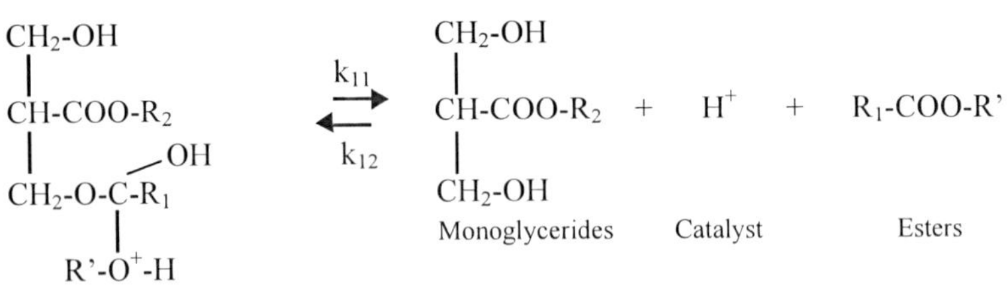

(10)

Monoglycerides to glycerol:

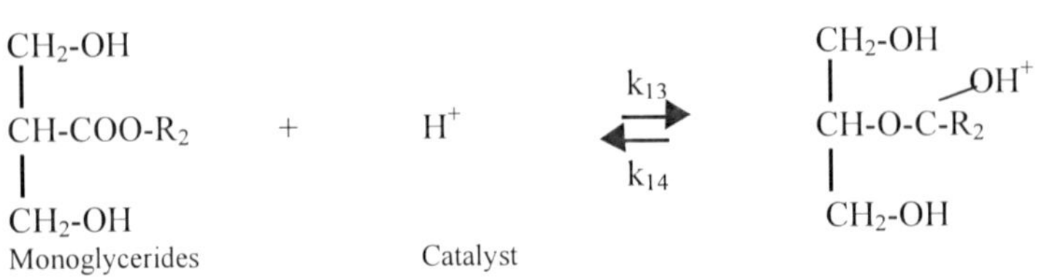

(11)

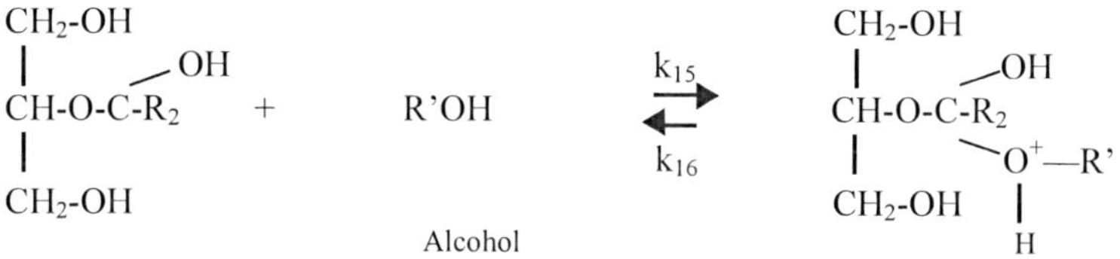

(12)

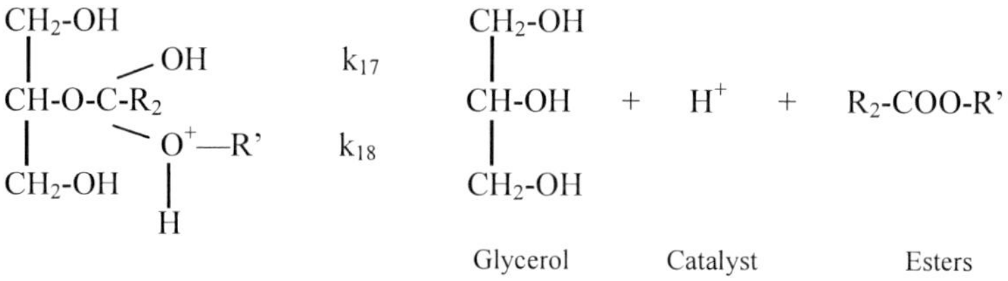

(13)

The previous mechanism shows the interaction of the catalyst (H^+) with the triglycerides as well as monoglycerides and diglycerides to produce ester and glycerol. Ester is produces in all three steps.

Many researchers have been working on the transesterification using sulfuric acid under different operational conditions as well as with different raw materials. The most standard case uses methanol, with a molar ratio of 32:1 related to the oil, a 1 or 2 % wt. of sulfuric acid, a reaction temperature of 60°C and a reaction time that could be over 30 hr [3,6-8]. However, many other operational conditions and raw materials have been tested. Among all the literature that could be found, some of them are presented here.

Goff et al. [28] have obtained full conversion of soybean oil transesterification in around 8 hr. reaction. They have used more drastic reaction temperature (120°C) and around 30:1 molar ratio of methanol. The final product has very small amount of free fatty acid. On the other hand, when the reaction temperature was around 100°C, it took over one day to obtain the same final condition. Simultaneously, the authors have tested several acids as catalyst, showing only good results for sulfuric acid. When nitric, hydrochloric, formic and acetic acid ere used, the final conversion was less than 0.7%. A similar work done by Zheng et al. [5] has showed that the mayor independent variables of the process were the molar ratio and the temperature. The higher molar ratio used in their work has make the reaction to be drove to completion in 4 h.

Morin et al. [29] have tested different Brønsted catalyst to carry on the transesterification reaction of rapeseed oil but using ethanol instead of methanol. Table 1 (extracted from reference [29]) show the selectivity for all the tested catalyst when T = 80°C, molar ratio 6:1, 500 rpm and 3 h of reaction.

Due to the fact that under some operational conditions, mass transfer limitations could become relevant, Ataya et al. [30] have studied the effects of mass transfer over the transesterification of refine oil and methanol using sulfuric acid as catalyst. The authors have found that it is possible to go from a two phase systems, to a single phase system by adding a solvent, like tetrahydrofuran. This is necessary to have a one phase system where there are no mass transfer limitations. The second alternative is to increase the stirring of the system, having a much better mix between the reactants as done by several authors. Other types of acid homogeneous catalyst have also being used for biodiesel production, such as Lewis acid catalyst. Di Serio et al [31] and Soriano et al. [32] have transesterified refine oils from different sources such as canola. It was found that this alternative could produce an over 98% conversion under a higher temperature (T=110°C), a similar molar ratio (24:1), a 5 times higher load of catalyst but in a shorter time (18hs). However, to obtain this result, a co-solvent was employed in order to produce a homogeneous phase.

Regarding the use of co solvent, Guan et al. [33] propose the use of another solvent, dimethyl ether, for their studied, the authors have chosen to use p-toluenesulfonic acid as catalyst, and have compared the system with and without co-solvent. Figure 5 (extracted from reference [33]) shows that for the studied temperature (40, 60 and 80°C) the use of co-solvent produces a net increase of 32% in the final yield (80°C case scenario).

In order to improve the use of high impure raw materials, Zullaikah et al. [34] have studied the transesterification of rice bran oil. During storage, rice bran could easily be changed by degradation (hydrolysis) into fatty acid, increasing the amount of it up to over 70%. In order to obtained biodiesel from this source, basic catalysts should be avoided; therefore, sulfuric acid appears as a good alternative. To obtain a biodiesel suitable for vehicle used, Zullaikah et al. [34] propose a two acid step catalysis process. During the first step, over 50% conversion was achieved at 60 °C consuming almost all the free fatty acid and over 30% of triglycerides. In the second step, with a higher temperature (100°C) the rest of the triglycerides were convert in less than 4 h and a final yield of over 98% was achieved. The authors have also showed that if the storage temperature is considerable reduce (up to 5°C), the amount of free fatty acid after 24 weeks is less than half compared when storage at room temperature, as seen in Figure 6. (Extracted from reference [34]).

Table 1. Comparison of ester selectivity for several catalyst [29]

Catalyst	TG conversion [%]	Ester selectivity [%]
$H_3PW_{12}O_{40}\_24H_2O$	27	95
$H_3PMo_{12}O_{40}\_28H_2O$	55	51
$H_4SiW_{12}O_{40}\_24H_2O$	20	44
$H_4SiMo_{12}O_{40}\_13H_2O$	45	63
H_2SO_4 (>95%)	27	62
H_3PO_4 (85%)	< 5	---

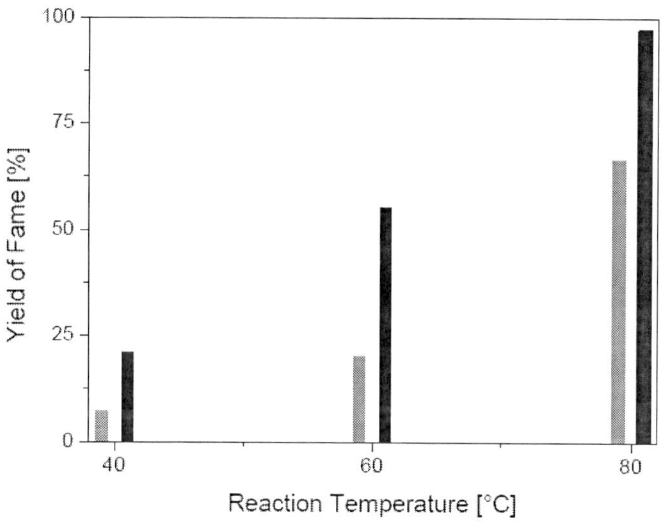

Figure 5. Comparison of biodiesel yield when dimethyl ether is used as co solvent. (■) no solvent, (■) with co-solvent [33].

Continuing with using waste vegetable oil, cooking oil is also possible to be modified into biodiesel, Wang et al. [35] have showed that it is possible to modified waste cooking oil with acid values of over 75 mg KOH/g with a single acid catalyst step. An evolution of the conversion into biodiesel over time is presented in Figure 7 (Extracted from reference [35]). It is also interested to know how each alcohol, or how different alcohols interact with the raw material and how the reaction is taking place. For that, methanol and ethanol, the most common used alcohol, should be tested and compared; Obibuzor et al. [36] have done some work related to this topic for waste oil with high amounts of free fatty acid. They have summarized their results in a table where the physical properties are showed and compared with the specifications from international standards while using ethanol and methanol for the production. Both cases were suitable showing a little lower value for the density of the ethyl ester. However, the final conversion was over 99% for both cases.

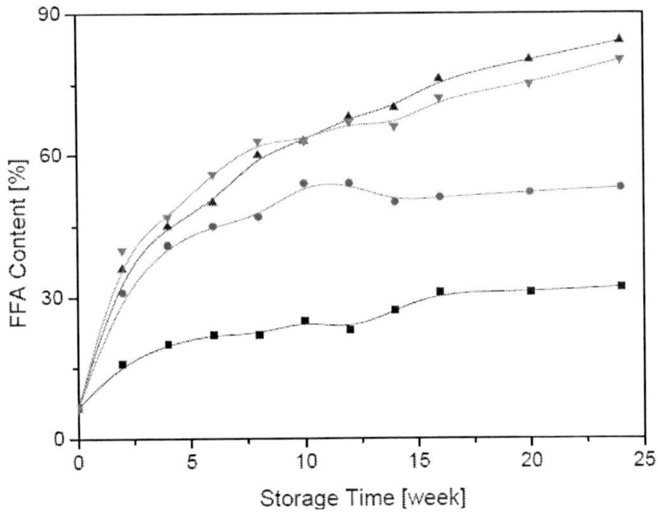

Figure 6. Storage temperature effect on FFA formation [34]. (■) dried sample, (●) Control, (▲) Control + 10 % water, (▼) Control + 20 % water.

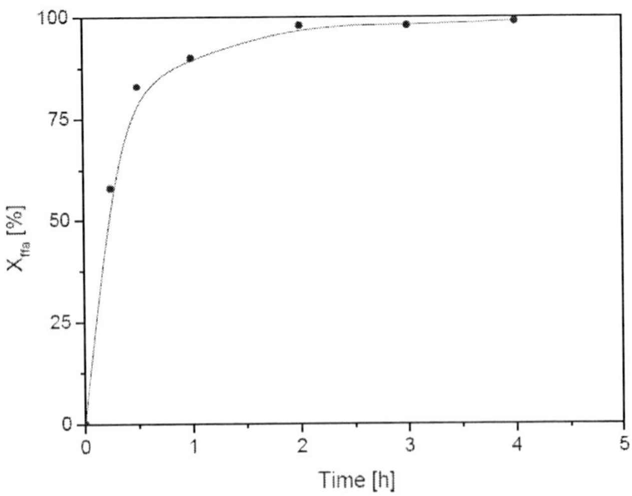

Figure 7. Conversion of FFA presented in a waste cooking oil [35].

Marchetti [9] have tested sulfuric acid as a main catalyst for the esterification as well as transesterification reaction of sunflower oil and ethanol to produce biodiesel. No co solvent was used since the agitation speed was high enough to assure one phase system. In this work, a lower molar ratio was used 6:1, with a temperature of 55°C and 1% wt of sulfuric acid. The initial mixture had 10% of free fatty acid in order to have an acid oil as raw material. The operational conditions here chosen are due to a possibility of comparison with the basic catalyst. Several operational conditions have been tested; modifications in all of them have changed the final conversion of the reaction rate. A summary of this is presented in Table 2.

Table 2. Operational conditions for the esterification reaction [9]

Case	Temp. [°C]	Molar ratio	Amount of catalyst[wt. %]	Initial FFA [wt. %]	Yield [%]
1	35	6:1	2	10	95
2	45	6:1	2	10	98
3	55	6:1	2	10	>99.5
4	45	4	2	10	95
5	45	10	2	10	>99.5
6	45	6:1	1	10	97
7	45	6:1	5	10	99
8	45	6:1	2	4	96
9	45	6:1	2	27	98

It was achieved a good final conversion for both the free fatty acid as well as for the triglycerides presented in the mixture. For the free fatty acid, as it was point out, it amount was reduce until 0.4% was reached, for the triglycerides the final conversion was high enough to assure a good fuel possible to be used in vehicles. However, the mayor disadvantage is the required time to achieve that result.

It has been also seen that this catalyst is suitable to carry on the transesterification reaction with good results, reaching the desirable conversion of over 98%. However the operational conditions are quite different from the conventional base technology. In this case the more common operational conditions are: T = 60°C (this is normally establish by the alcohol employed, due to its boiling point), alcohol: methanol, molar ratio of alcohol/oil = 32, reaction time is 40 h and an amount of catalyst is 1% w/w. With these operational conditions, the final biodiesel has good quality and should be able to be used in regular engines. Even more, this technology does allow the esterification reaction as well as the transesterification reaction to take place, which allow us to use it for a big variety of raw materials with different amount of impurities such as free fatty acids, water, etc. with no risk of soap formation and no need of physical refining.

Due to the relevance of the process a short summary of kinetics expression for both reaction it is introduce.

4. KINETICS

When using an acid oil, both reactions could take place simultaneously; therefore, several studies have been done in relation to the esterification reaction kinetic and to the transesterification reaction kinetics.

A short summary for both reactions is presented below.

Since both reactions are taking place simultaneously in the reactor and could take some place simultaneously in time, as seen in lab work, two different models should be presented here.

1. Triglyceride does not interact in the reaction.
2. Triglycerides interact in the reaction.

4.1. Kinetic Model for the Esterification Reaction

The esterification reaction with sulfuric acid has been widely studied for fatty acid as raw materials. Zhou et al. [15] analyzed the esterification of oleic acid in the presence of methanol while using toluene sulfonic as catalyst. For this scenario, a second reversible order reaction is proposed. On their work, Goto et al. [16] have studied the same reaction but with palmitic acid while using iso-alcohol and sulfuric acid as catalyst. They also propose a second order reversible kinetic expression. The same kinetic model was obtained for Bart et al. [12] experimental data. In the latest work, levulinic acid was used with sulfuric acid and butanol.

The kinetic mechanism for the reaction was established by Straiweiser et al. [37]. This idea was also employed by Ronnback et al. [14] who have used this mechanism for the esterification of acetic acid and methanol. In that work, the authors have also introduced the possibility that the mixture is not an ideal one, however, they reach the result that there was no difference between ideal and non ideal mixture.

Several expression were found for the fatty acid in the presence of triglycerides, Sendzikiene et al. [20] as well s Berrios et al. [19] have proposed kinetics expression for fatty acid and methanol in a triglycerides environment.

Both works proposed a first order advance kinetic, while Berrios et al. [19] also proposed a second order reverse reaction. However, no comments were introduced related to the triglycerides.

To obtain the kinetic expression when the triglycerides behave like inert, the esterification reaction mechanism developed by Straiweiser et al. [37] was used.

The steps involved are the following ones:

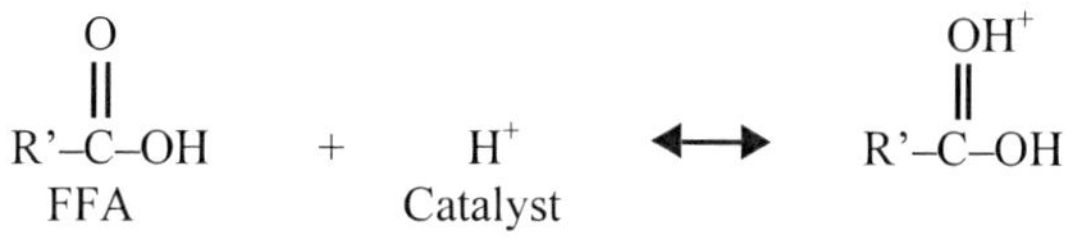

(14)

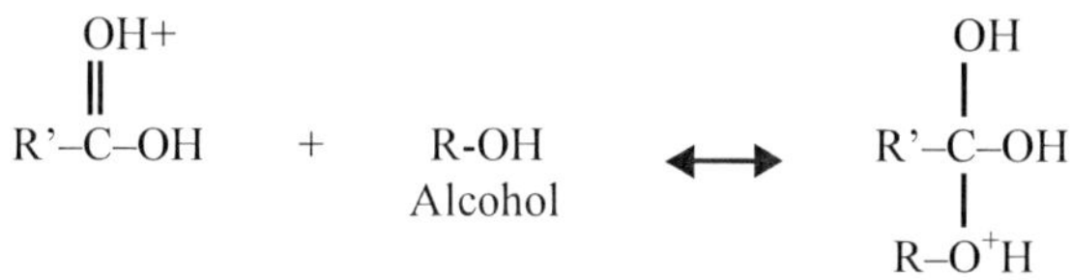

(15)

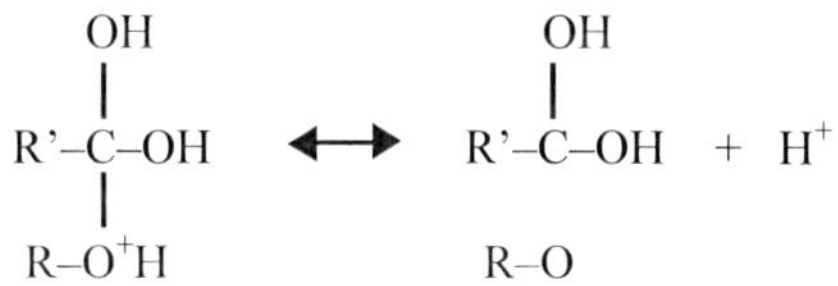

(16)

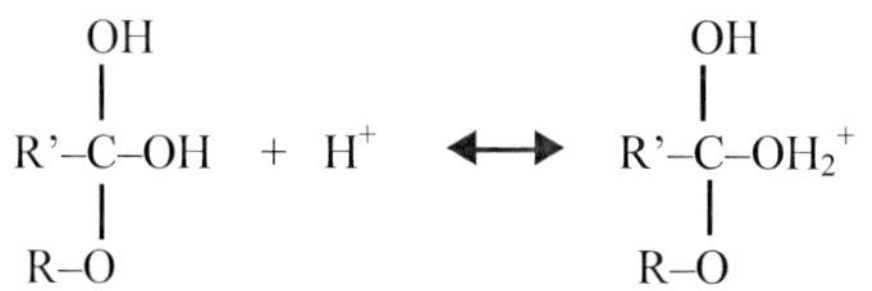

(17)

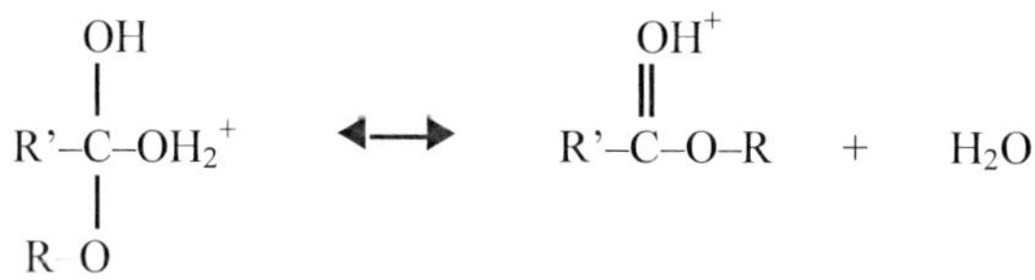

(18)

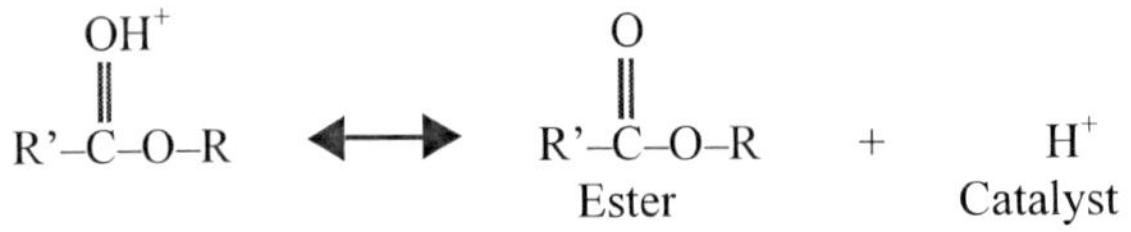

(19)

From the previous mechanism, step 14 was considered the show step when applying the controlling step method. This was a result of studying all the possible controlling steps.

The final reaction rate expressions as well as the catalyst concentration are as follows:

$$r_{ffa} = m_{cat} k \left(C_{ffa} - \frac{C_{faee} C_w}{C_{alc} K} \right) \tag{20}$$

$$m_{cat} = \frac{m^0_{cat}}{1 + K_1 C_{faee} C_w + \frac{K_2 C_{faee} C_w}{C_{ffa}} + K_3 C_{faee}} \tag{21}$$

where $k = k_\infty \exp^{\left(-\frac{E}{RT}\right)}$ y $K = K_\infty \exp^{\left(-\frac{\Delta H}{RT}\right)}$

Using the experimental data presented by Marchetti and Errazu [21] the kinetics parameters have been obtained (Table 3.)

An activation energy of 5530 cal/mol, which is close the one obtained by Sendzikiene et al. [20] (3176.65 cal/mol) was fitted.

In Figure 8 and 9 it could be seen how this model matches the experimental data for several temperatures as well as different amount of catalyst. As it can be seen from the figure, the model fits considerable well the experimental data. These figures were obtained from reference [9].

In Figure 9 it can be observed how the model approaches the experimental data when different amounts of catalyst are being used. The kinetic model has great accuracy with the data. The mayor error appears for smallest amount of catalyst due to the possible experimental error.

Despite the good agreement between the experimental data and the mathematical model, the presence of triglycerides might have an effect over the general model for both reactions. Because of this, a new kinetic expression was developed for both, esterification and transesterification reaction, simultaneously.

Table 3. Kinetics parameters values

Parameter	Value	Units
k_∞	2.11 10^2	l/(kg hr)
E	5.53 10^3	cal/mol
K_∞	6.42 10^{67}	
ΔH	9.55 10^4	cal/mol
K_1	7.62 10^{-2}	l^2/mol^2
K_2	3.14 10^1	l/mol
K_3	1.00 10^{-11}	l/mol

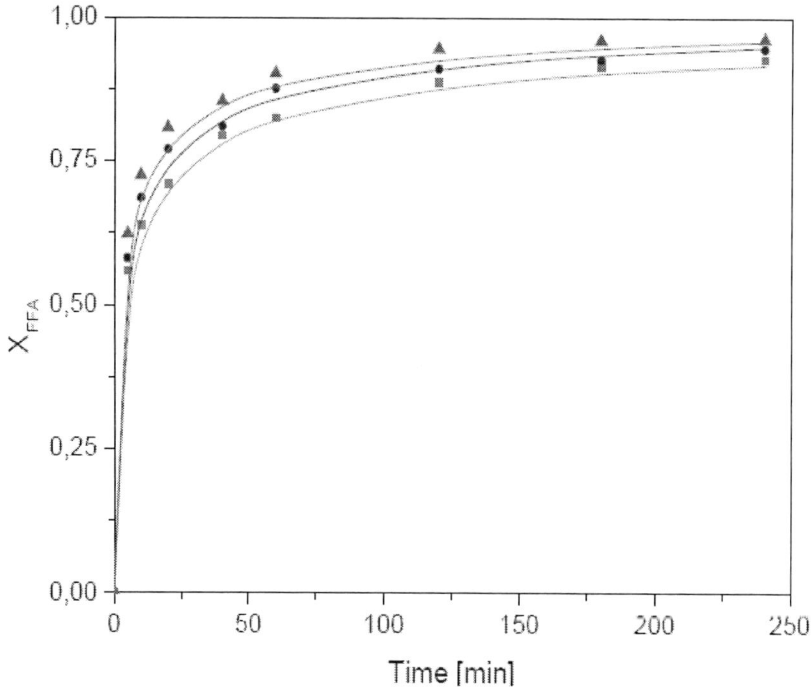

Figure. 8. Conversion of free fatty acid. T = 35°C. (■) Exp. (—) Model T = 45°C. (●) Exp. (—) Model T = 55°C. (▲) Exp. (—) Model.

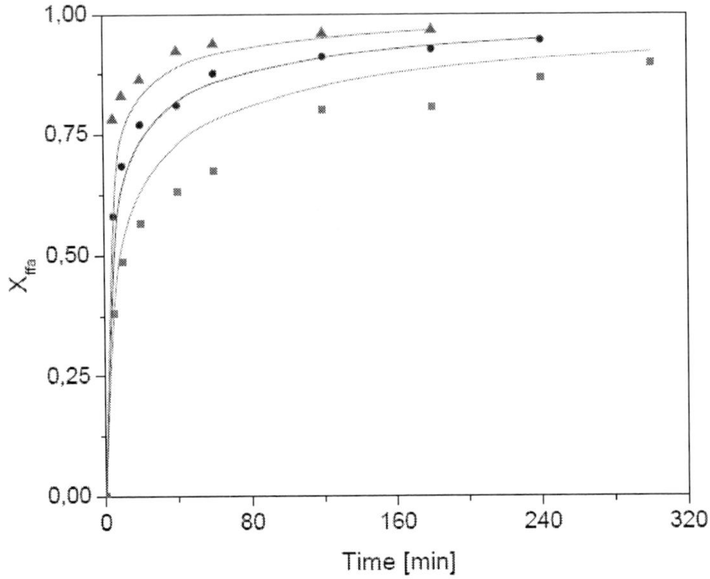

Figure 9. Conversion of FFA. S = 1.03% (■) Exp. (–) Model. S = 2.26% (●) Exp. (–) Model. S = 5.14% (▲) Exp. (–) Model.

4.2. Kinetic Model for Both Reactions Simultaneously

To obtain a more general expression for both reactions simultaneously, it is important to establish which are the steps involved, as are showed, for the esterification, the mechanism proposed by Straiweiser et al. [37] seems accurate and generally used. For the transesterification reaction, even there is a mechanism for the interaction of the catalyst with the vegetable oil, the most common kinetic mechanism is based in equation (1 to 3), considering each of them as elementary steps.

The simplified reactor scheme, highly used, is as follows.

$$TG + ALC \leftrightarrow DG + FAEE \quad (22)$$

$$DG + ALC \leftrightarrow MG + FAEE \quad (23)$$

$$MG + ALC \leftrightarrow G + FAEE \quad (24)$$

$$FFA + ALC \leftrightarrow H_2O + FAEE \quad (25)$$

Considering the reaction scheme, the mass balance for each component in the batch reactor is:

$$\frac{dC_{tg}}{dt} = -r_1 \qquad \frac{dC_{dg}}{dt} = r_1 - r_2$$

$$\frac{dC_{mg}}{dt} = r_2 - r_3 \quad \frac{dC_{g}}{dt} = r_3 \qquad \frac{dC_{faee}}{dt} = r_1 + r_2 + r_3 + r_4$$

$$\frac{dC_{alc}}{dt} = -\frac{dC_{faee}}{dt} \qquad \frac{dC_{agl}}{dt} = -r_4$$

where r_i ($i = 1,4$) are the reaction rate for reactions 22 a 25).

Even when both reactions are considered to be simultaneously, it is possible to have some independency in both expressions, because of this, the esterification mechanism developed by Straisweiser et al. [37] will continue to be applied, adding a mechanism for the transesterification reaction. A power law mechanism was used, obtaining the following expressions.

$$r_1 = k_1 C_{tg} C_{alc} - k_2 C_{dg} C_{est} \tag{26}$$

$$r_2 = k_3 C_{dg} C_{alc} - k_4 C_{mg} C_{est} \tag{27}$$

$$r_3 = k_5 C_{mg} C_{alc} - k_6 C_{g} C_{est} \tag{28}$$

While the expression for the esterification remains as follows

$$r_4 = \frac{m_{cat}^0}{1 + K_1 C_{est} C_w + \frac{K_2 C_{est} C_w}{C_{agl}} + K_3 C_{est}} \left(k_7 C_{agl} - \frac{k_8 C_{est} C_w}{C_{alc}} \right) \tag{29}$$

And the kinetic constants follow the Arrhenius form:

$$k_i = k_{i_\infty} \exp^{\left(-\frac{E}{RT}\right)} \qquad i = 1,8 \tag{30}$$

All the above equations are use in the model of the reactor and will predict the evolution and distribution of reactants as well as products for the time were the reaction takes places.

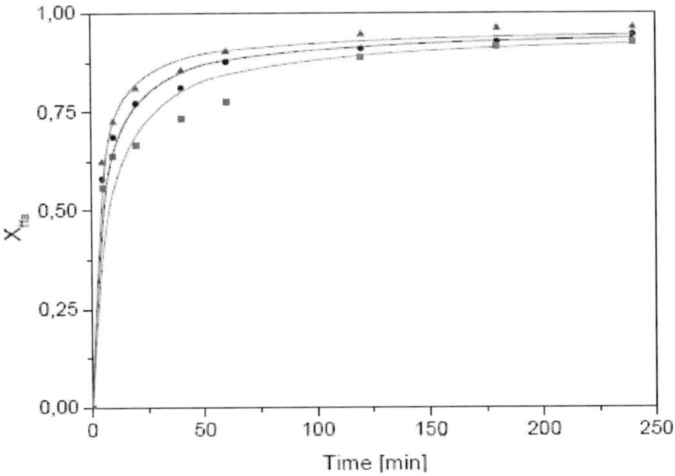

Figure. 10. Free fatty acid conversion. T = 35°C. (■) Exp. (—) Model, T = 45°C. (●) Exp. (—) Model, T = 55°C. (▲) Exp. (—) Model.

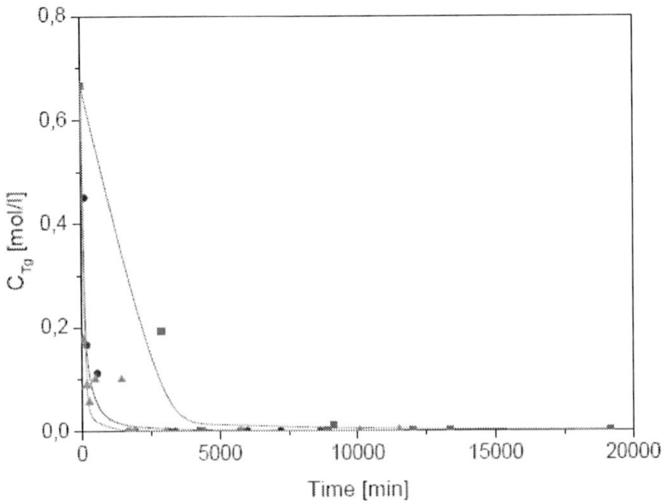

Figure 11. Triglycerides concentration vs. time. T = 35°C. (■) Experimental, (—) Model T = 45°C. (●) Experimental, (—) Model, T = 55°C. (▲) Experimental, (—) Model.

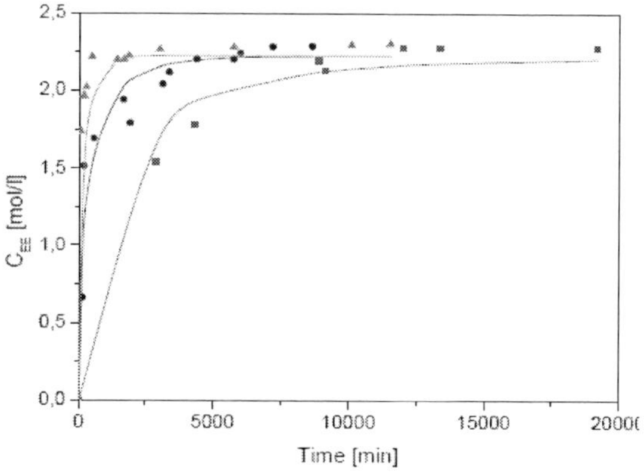

Figure 12. Esters concentration vs. time. T = 35°C. (■) Experimental, (—) Model, T = 45°C. (●) Experimental, (—) Model, T = 55°C. (▲) Experimental, (—) Model.

Table 4. Kinetics parameters

Parameter	Value	Units
$k_{\infty 1}$	5.73 10^3	l/(mol h)
E_1	9.25 10^3	cal/mol
$k_{\infty 2}$	1.18 10^3	l/(mol h)
E_2	9.25 10^3	cal/mol
$k_{\infty 3}$	1.15 10^4	l/(mol h)
E_3	9.25 10^3	cal/mol
$k_{\infty 4}$	2.15 10^4	l/(mol h)
E_4	9.25 10^3	cal/mol
$k_{\infty 5}$	3.66 10^{14}	l/(mol h)
E_5	2.56 10^4	cal/mol
$k_{\infty 6}$	1.45 10^0	l/(mol h)
E_6	9.26 10^3	cal/mol
$k_{\infty 7}$	3.42 10^4	l/(kg h)
E_7	9.08 10^3	cal/mol
$k_{\infty 8}$	6.00 10^{-2}	l/(kg h)
E_8	3.05 10^3	cal/mol
K_1	1.17	l^2/mol^2
K_2	1.10	l^2/mol^2
K_3	2.00	l/mol

In Figure 10 it can be seen how the fatty acid conversion evolved and how good the mathematical modeling is. In Figure 11 and Figure 12 it can be seen the evolution of the concentration of triglycerides as well as ester. The agreement is good for all the tested temperatures, showing that considering both reactions simultaneously give a good accuracy.

Once the model was converged and the experimental data satisfied, the kinetics parameter were obtained and check about the physical meaning of their values, in order to see if the result was physically possible. Table 4 shows these results.

Since two different models were tested considering different physical scenarios, it will be important to verify that the case that considered both expressions is actually more accurate than the case with the triglycerides as inert.

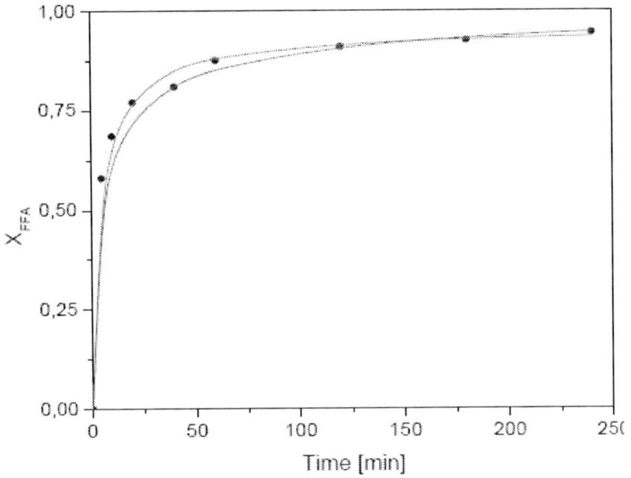

Figure 13. Conversion of free fatty acids. (—) Esterification, (—) esterification + transesterification, (●) Experimental data.

This should be the situation due to the fact that both equations have some overlapping in time. Figure 13 shows how both model approach the experimental data for a fixed operational conditions, It could be seen that the model with both reactions represents better the experimental results. This is something to be expected because the model is more a like to what is happening in the reaction medium

4.3. More Complex Mechanism

When we look at the mechanism at the beginning of this chapter and how the catalyst interacts with the triglycerides, it is easy to see that the predictions in the above section is somehow incomplete, even though it does represent the experimental data, some more complex mechanism, that will considered the transesterification not as elementary step but as a sum of elementary steps could be proposed. For this, the mechanism proposed by Straiweiser et al. [37] was used one more time for the esterification reaction, while for the transesterification it was employed the mechanism proposed by López et al. [26] and by Schuchardt et al. [27].

After considering all possible mechanism, and test them against experimental data it was found that the expression for each of the reaction rate would take the following form

$$r_1 = \left(k_5^{TE} K_{1234}^{TE} C_{tg} C_{alc} - k_6^{TE} C_{faee} C_{dg}\right) * m_{cat} \tag{31}$$

$$r_2 = \left(k_{11}^{TE} K_{78910}^{TE} C_{dg} C_{alc} - k_{12}^{TE} C_{faee} C_{mg}\right) * m_{cat} \tag{32}$$

$$r_3 = \left(k_{17}^{TE} K_{13141516}^{TE} C_{mg} C_{alc} - k_{18}^{TE} C_{faee} C_{g}\right) * m_{cat} \tag{33}$$

$$r_4 = \left(k_3^{ED} C_{agl} C_{alc} K_{12}^{ED} - \frac{k_4^{ED} C_{faee} C_w}{K_{3456}^{ED}}\right) * m_{cat} \tag{34}$$

$$m_{cat} = \frac{m_{cat}^0}{\left(\begin{array}{l} 1 + C_{agl} K_1^{ED} + \dfrac{C_{faee} C_{agua}}{K_{3456}^{ED}} + \dfrac{C_{faee} C_{agua}}{K_{56}^{ED}} + \dfrac{C_{faee}}{K_6^{ED}} + C_{tg} K_{12}^{TE} + \\ C_{tg} C_{alc} K_{1234}^{TE} + C_{dg} K_{78}^{TE} + C_{dg} C_{alc} K_{78910}^{TE} + C_{mg} K_{1314}^{TE} + C_{mg} C_{alc} K_{13141516}^{TE} \end{array}\right)} \tag{35}$$

where K_{ij} correspond to the equilibrium constant obtain from the advance constant k_i and the reverse one k_j. K_{ijwk} is the result of considering K_{ij} and K_{wk} together and multiplying each other. The upper index is related to the reaction they are coming from, ED is for direct esterification and TE is for transesterification.

Since this is a first step and it is quite more complex than the previous one, the study must continue in order to obtain a better kinetic expression, however, as a first result, Figure 14 shows the simulation and the experimental data. In the plot it was added line showing the error bars set at 15%.

Figure 14 shows a good agreement for triglycerides, esters and fatty acid, with the higher deviation in monoglycerides as well as in diglycerides.

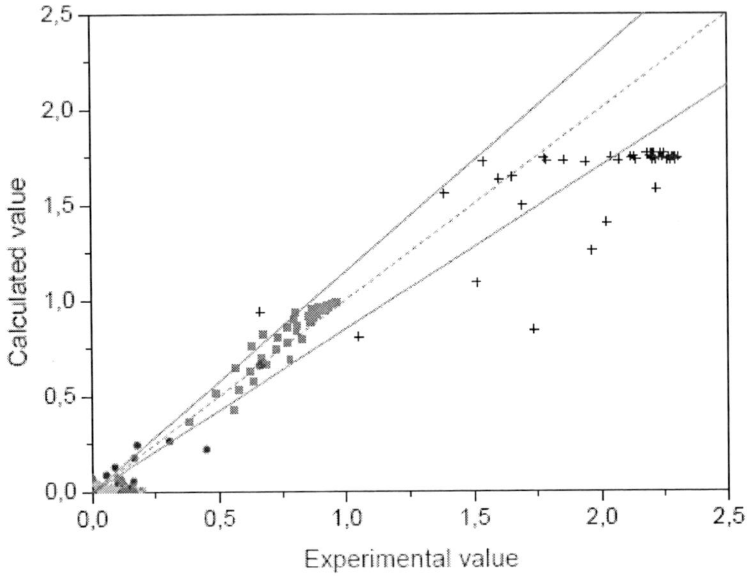

Figure 14. Comparison of experimental and calculated values. (■) FFA, (●) TG, (▲) DG, (×) MG and (+) FAEE.

This model shows a good correlation between the calculated values for all the compounds involved in the two main reactions. However, a much more in detail optimization is needed in order to assure that the complexity of this model, and the new reaction mechanism, is in better accordance to the experimental values than the simpler previous approach.

On the other hand, more reactions could also be considered, the possible interaction of TG with water to produce esters or the interaction of triglycerides with glycerol to produce diglycerides and monoglycerides. However, a more detail study if these operational conditions and catalysts employees are suitable to allow those reactions to happen is necessary.

5. ECONOMIC STUDY

As previously showed for the homogeneous technology, economic studies are also relevant when analyzing new technological alternatives such as sulfuric acid.

Considering the higher reaction time involved in this alternative, and the fact that it will be possible to treat less pure raw material, a direct comparison with the conventional process is not possible. Nevertheless, researches have been done in this matter to see the feasibility of this technology, such as those carried on by Zhang et al. [1], Marchetti [9] and Marchetti et al. [10-11]. While Zhang et al. [1] showed that is economical possible to use this option to produce technology, Marchetti [9] have showed that it is possible if the internal rate for the net present value is not too high. If this is over 7% then it might become not economical viable. However, one very important variable not discuss here and that should be considered very carefully is the how the process is in itself, meaning, how many equipments for reaction, is there separation between them, what type of purification is being used. All these considerations/assumptions are extremely important when making an economic analysis, since a different scenarios will most likely, have a different outcome result.

REFERENCES

[1] Zhang, Y., Dubé, M.A., McLean, D.D., Kates, M. *Biodiesel production from waste cooking oil: 2 Economic assessment and sensitivity analysis. Bioresource Technology*. 90. (2003), 229-240.

[2] Haas, M.J., McAloon, A.J., Yee, W.C., Foglia, T.A. *A process model to estimate biodiesel production costs. Bioresource Technology*. 97(4). (2006), 671-678.

[3] Marchetti, J.M., Miguel, V.U., Errazu, A.F. *Possible methods for biodiesel production*. Renewable and Sustainable Energy Reviews. 11. (2007), 1300-1311.

[4] Knothe, G., Van Gerpen, J., Krahl, J. (Eds.), *The biodiesel handbook*, AOCS Press, Champaign, Illinois, 2005, 302 pp.

[5] Zheng, S., Kates, M., Dubé, M.A., McLean, D.D. *Acid-catalyzed production of biodiesel form waste frying oil. Biomass and Bioenergy*. 30(3). (2006), 267–272.

[6] Canakci M, Van Gerpen J. *Biodiesel production via acid catalysis*. Transactions of the ASAE. 42(5). (2003), 1203–1210.

[7] Canakci M, Van Gerpen J. *Biodiesel production from oils and fats with high free fatty acids*. Transactions of the ASAE. 44(6). (2003), 1429–1436.

[8] Van Gerpen, J. *Biodiesel processing and production. Fuel Processing Technology*. 86. (2005), 1097-1107.

[9] Marchetti, J.M. Ph.D. thesis. *Technological alternatives for biodiesel production*. Universidad Nacional del Sur. Argentina (2008).

[10] Marchetti, J.M., Errazu, A.F. *Technoeconomic study of supercritical biodiesel production plant. Energy Conversion and Management*. 49. (2008), 2160-2164.

[11] Marchetti, J.M., Miguel, V.U., Errazu, A.F. *Techno-economic study of different alternatives for biodiesel production. Fuel Processing Technology*. 89. (2008), 740–748.

[12] Bart, H.J., Reidetschläger, J., Schatka, K. Lehmann, A., *Kinetics of esterification of levulinic acid with n-butanol by homogeneous catalysis*. Industrial and Engineering Chemistry Research. 33. (1994), 21-25.

[13] Grob, S., Hasse, H. *Reaction Kinetics of homogeneously catalyzed esterification of 1-butanol with acetic acid in a wide range of initial compositions. Industrial and Engineering Chemistry Research*. 45. (2006), 1869-1874.

[14] Rönnback, R., Salmi, ., Vuori, A., Haario, H., Lehtonen, J., Sundqvist, A., Torronen, E. *Development of a kinetic model for the esterification of acetic acid with methanol in the presence of a homogeneous acid catalyst. Chemical Engineering Science*. 52(19). (1997), 3369-3381.

[15] Zhou, M., Gilot, B., Domenech, S. *Modelisation d'un réacteur d'estérification. Partie 1 : Estérification de l'acide oléique par le méthanol. Détermination des données thermodynamiques et cinétiques. Entropie* 120. (1984), 3-10.

[16] Goto, S., Tagawa, T., Yusoff, A. *Kinetics of the esterification of palmitic acid with isobutyl alcohol. International Journal of Chemical Kinetics*. 23. (1991), 17-26.

[17] Liu, Y., Lotero, E., Goodwin Jr. J.G. *A comparison of the esterification of acetic acid with methanol using heterogeneous versus homogeneous acid catalysis. Journal of Catalysis*. 242. (2006), 278-286.

[18] Chen, X., Xu, Z., Okuhara, T. *Liquid phase esterification of acrylic acid with 1-butanol catalyzed by solid acid catalyst*. Applied Catalysis A: General. 180. (1999), 261-269.

[19] Berrios, M., Siles, J., Martín, M.A., Martín, A. *A kinetic study of the esterification of free fatty acids (FFA) in sunflower oil. Fuel* 86. (2007), 2383-2388.

[20] Sendzikiene, E., Makareviciene, V., Janulis, P., Kitrys, S. *Kinetics of free fatty acids esterification with methanol in the production of biodiesel fuel. European Journal of Lipid Science and Technology*. 106. (2004), 831-836.

[21] Marchetti, J.M., Errazu, A.F. *Esterification of free fatty acids using sulfuric acid as catalysts in the presence of triglycerides. Biomass and Bioenergy*. 32. (2008), 892-895.

[22] Ma, F. Hanna, M. A. *Biodiesel production: a review. Bioresource Technology*. 70. (1999), 1-15.

[23] Fukuda, H. Kondo, A. Noda, H. Biodiesel *Fuel Production by Transesterification of Oils. J. Bioscience and Bioengineering*. 92(5). (2001) 405-416.

[24] Barnwal, B.K. Sharma, M.P. *Prospects of Biodiesel production from vegetable oils in India. Renewable and Sustainable Energy Reviews*. 9(4). (2005) 363-378

[25] Srivastava, A. Prasad, R. *Triglycerides-based diesel fuels. Renewable and Sustainable Energy Reviews*. 4. (2000) 111-133.

[26] Lopez, D.E., Goodwing Jr. J.G., Bruce, D.A., Lotero, E. *Transesterification of triacetin with methanol on solid acid and base catalysts*. Applied Catálisis A: General 295. (2005), 97-105.

[27] Schuchardt, U., Serchelim R., Vargas, R.M. *Transesterification of vegetable oils: a review. Journal of Brazilian Chemist Society*. 9. (1998), 199-210.

[28] Goff, M.J., Bauer, N.S., Lopes, S., Sutterlin, W.R., Suppes, G.J. *Acid-catalyzed alcoholisis of soybean oil. Journal of the American Oil Chemistry Society*. 81(4). (2004), 415-420.

[29] Morin, P., Hamad, B., Sapaly, G., Carneiro Rocha, M.G., Pries de Oliveira, P.G., Gonzalez, W.A., Andrade Sales, E., Essayem, N. *Transesterification of rapeseed oil with ethanol: I. Catalysis with homogeneous keggin heteropolyacids*. Applied Catalysis A: General. 330(10). (2007), 69-76.

[30] Ataya, F., Dubé, M.A., Ternan, M. *Acid-Catalyzed Transesterification of Canola Oil to Biodiesel under single and two phase reaction conditions. Energy and Fuels*. 21. (2007), 2450-2459.

[31] Di Serio, M., Tesser, R., Dimiccoli, M., Cammarota, F., Nastasi, M., Santacesaria, E. *Synthesis of biodiesel via homogeneous Lewis acid*

catalyst. Journal of Molecular Catalysis A: Chemical. 239. (2005), 111-115.

[32] Soriano, N.U., Venditti, R., Argyropoulos, D.S. *Biodiesel synthesis via homogeneous Lewis Acid Catalyzed transesterification. Fuel.* 88. (2009), 560-565.

[33] Guan, G., Kusakabe, K., Sakurai, N., Moriyama, K. *Transesterification of vegetable oil to biodiesel fuel using acid catalysts in the presence of dimethyl ether. Fuel.* 88. (2009), 81-86.

[34] Zullaikah, S., Lai, C.C., Vali, S.R., Ju, Y.H. *A two step acid catalyzed process for the production of biodiesel from rice bran oil. Bioresource Technology.* 96(17). (2005), 1889-1896.

[35] Wang, Y., Ou, S., Liu, P., Xue, F., Tang, S. *Comparison of two different processes to synthesize biodiesel by waste cooking oil. Journal of Molecular Catalysis A: Chemical.* 252. (2006), 107-112.

[36] Obibuzor, J.U., Abigor, R.D., Okiy, D.A. *Recovery of oil via acid catalyzed transesterification. Journal of the American Oil Chemistry Society.* 80(1). (2003), 77-80.

[37] Streitwieser. A., Heathcock. C.H., Kosower. E.M. *Introduction to organic chemistry. Fourth Edition.* ISBN: 0-02-418170-6.

Chapter 6

HETEROGENEOUS CATALYSIS

1. INTRODUCTION

The use of conventional homogeneous technology has several advantages such as short reactions time as well as low temperature; however, it has one important drawback related to the purity of the raw material needed as well as to the purification of the products and the downstreaming separation of possible effluents.

If basic alternative is used, as mentioned before, the oil has to be refine to avoid soap formation; however, even if the oil is refine the use of an homogenous catalyst produced a demand on neutralization as well as purification and separation, increasing the cost of the process and producing more effluents to be treated. A way to avoid this could be obtained by using a heterogeneous catalyst. There is one more advantage if the catalyst allows it, the possibility to carry on the esterification reaction as well as the transesterification. If both are taking place simultaneously, the needs of a pre esterification step or of a purification section in the production plan will not be required.

Even more, a solid catalyst could be used in a fix bed reactor, as showed in Figure 1 allowing the process to achieve a better quality of the main products, reducing the amount of equipment in the plant, having a less initial investment and producing a less amount of effluents and pollutants.

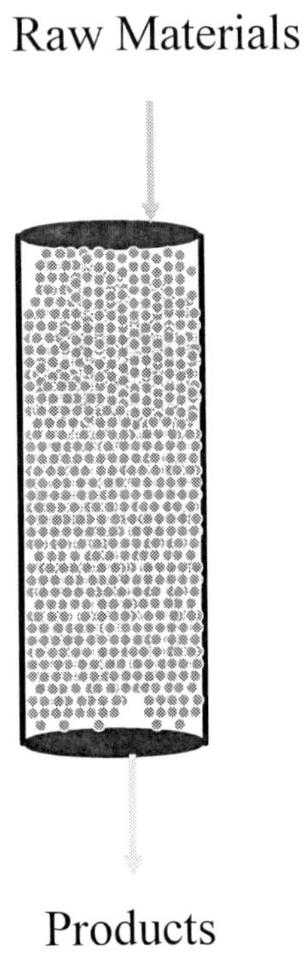

Figure 1. Schematic representation of a fix bed reactor.

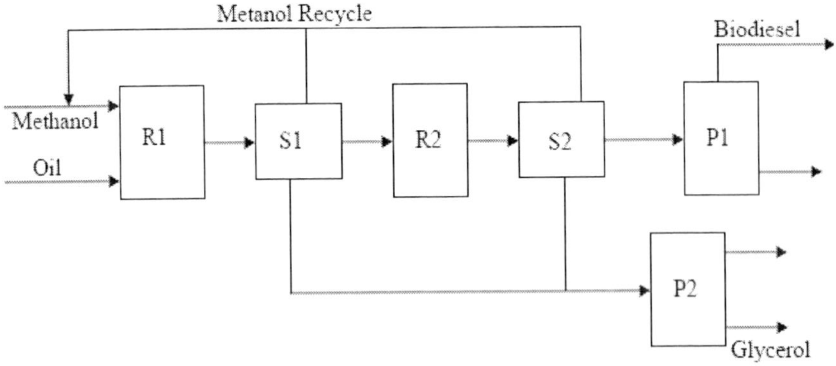

Figure 2. Block diagram of a heterogeneous production plant. R1 and R2 are the reactors of the system, S1 and S2 are the separation process while P1 and P2 are the purification area of the products.

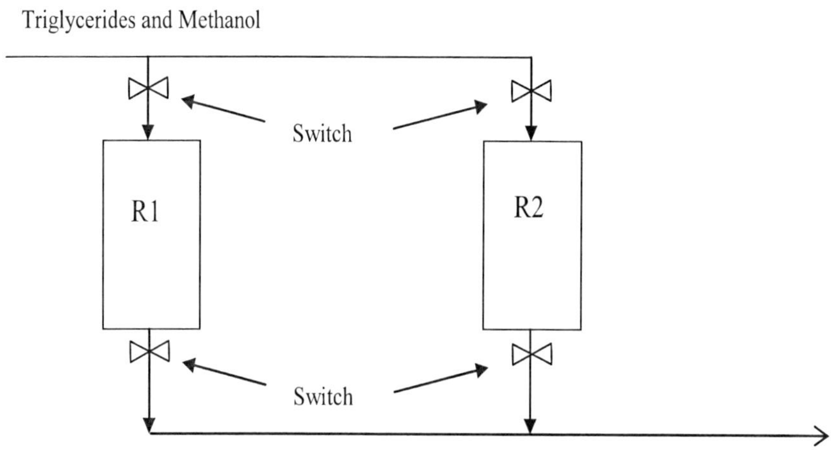

Figure 3. Possible system for reaction and regeneration of a heterogeneous catalyst.

Moreover, heterogeneous catalyst that allows several amounts of impurities are the key factor to be used when high impure raw materials are used such as frying oil, waste oils or even soapstocks where the amount of free fatty acid could be higher than 60% [1]. A schematic design of a complete production plant is presented in Figure 2. It has two reaction steps with separation among them in order to improve further purification. A simple distillation could separate the methanol form the other phases which will be further separated in another equipment. The purity of all the products is quite high due to the fact that there is no presence of external washing/purifications compounds [2].

However, the solid resin might need purification and regeneration in order to carry on the process in a continuous way. A schematic of a possible design for a continuous reaction/regeneration process is presented in Figure 3 [3]. It is important to point out that not all the process required this and not all the systems are continuous, some technologies could also work as batch process with great results and no need of regeneration of the catalyst [4].

In the following sections it will be discuses the esterification reaction as well as the transesterification reaction in order to provide a comparative study with the previous and future technologies described in this book. Both reaction will be presented separately first to then be presented together with a kinetic study and a short economic analysis.

Table 1. Literature review for the esterification reaction

Fatty acid	Alcohol	% Catalyst	Temp. [°C]	x	Reaction time	Molar ratio	Ref.
Acetic acid	1-Butanol	1 a 20 g/gmol Amberlyst-15 Amberlite IR-120 Indion-130 TX66	70-90	30-84%	92 h. aprox.	3:1	[6]
Acetic acid o	1-Butanol	1.9g Amberlyst 15 1.9g Smopex-101 5g Sulfated Zirconium 5.07g Niobic acid	75	15-75	350 min.	1:1	[7]
Acetic acid	Amyl	7.24 g Amberlyst 15	30-80	64-96%	8000 [g min. cm^{-3}]	1:1 to 10:1	[8]
Acetic acid o	Amyl	Dowex 50Wx8-100	50-120	64.6-96.4	0.6-10.93 [g min. cm^{-3}]	1:1 to 10:1	[9]
Acetic acid	Ethanol	Amberlyst 15	40-70	60-99%	300 min.	0.25:1 a 6:1	[10]
Acetic acid o	Ethanol	Smopex-101	60	20-85%	500 min.	1:4 to 4:1	[11]
Acetic acid	Iso-Amyl	2.5 a 10 wt% Purolite C-175	60-90	75-85%	250-400 min.	2:1 to 10:1	[12]
Acetic acid	Iso-butanol	0.775 g/l Amberlite IR120	45	10-45%	80-700 h.	1:1	[13]
Acetic acid	Methanol	1.09g/45ml Nafion/silica	60	75%	650 min.	2:1	[14]
Acetic acid	Methanol	1g Zirconium tungsten	60	----	2 h.	2:1	[15]
Acetic acid	Methanol	3.81-59.68g Amberlyst 15	30-70	10-95%	1.3-95 h.	0.62:1 to 19.35:1	[16]
Acetic acid	Methanol Ethanol 1-Propanol 2-Propanol 1-Butanol	Smopex-101	60	46-65%	500-3000 min.	1:1	[11]
Acrylic acid	1-butanol	0.25g Amberlyst 15 1g $Cs_{2.5}H_{0.5}PW_{12}O_{40}$ 1g SO_4^{-2}/ZrO_2 1g Nafion-H	80	14-60	4 h.	1:1	[17]

Fatty acid	Alcohol	% Catalyst	Temp. [ºC]	x	Reaction time	Molar ratio	Ref.
Lactic acid	Ethanol	62.7g Amberlyst XN-1010	95	70%	480 min.	2.348:1	[18]
Palmitic acid	Iso-butanol	0.68 a 5.31 g Amberlyst 15	107	20-60	270 min.	2.5:1 to 10:1	[19]
Lactic acid	Ethanol	Amberlyst-15 D001 4% (w/w) D002 4% (w/w) NKC 2-6% (w/w) 002	60-88	25% aprox.	450 min.	3:1	[20]
Lactic acid	Ethanol	1.1 a 6 wt% Amberlyst 15	55-85	38-85%	1600 min.	1:1 to 15:1	[21]
Lactic acid	Methanol	1.1 a 6% (w/w) Dowex 50W8x Dowex 50W2x Amberlyst 36 Amberlyst 15	42-80	8-35 %	250	0.5:1 to 5:1	[22]
Maleic acid	Ethanol	50-150 kg/m^3 Indion-170 Amberlyst 36 Amberlyst15 Amberlite IRA 120 20%DTP/K-10	50-80	28-78	210 min.	3:1 to 15:1	[23]
Oleic acid	1-butanol	4.9-9.25% Amberlite IR-100	114-116	45-98%	12 h.	4:1	[24]
Oleic acid	Octadecenol	0-0.039 mol/l cobalt chloride	116-164	20-98%	550 min.	0.66:1 to 2.66:1	[25]
Oleic acid	Methanol	10 % HMS-SO_3H SBA-15- SO_3H SBA-15-ph- SO_3H	85-120	50-90%	180 min.	20:1	[26]
Mix of fatty acid	Methanol	1 g. HMFI(25) HMOR	60	80%	3 h	30:1	[27]

Table 1. Some Works done on esterification reaction (Continued)

Fatty acid	Alcohol	% Catalyst	Temp. [ºC]	x	Reaction time	Molar ratio	Ref.
Propanoic. acid	Methanol Ethanol 1-propanol 2-propanol 1-butanol	Smopex-101	60	18-68%	500-3000 min.	1:1	[11]
Palmitic acid	Methanol Ethanol	0.2 g of D50w8 D50w2 PVA_SSA20 PVA_SSA40	60	75-94%	2-24 h	90	[27]
Propionic acid	1-butanol	1 wt% Amberlyst 15 Amberlyst 35 Amberlyst 39 HZSM-5	80-110	38-50%	6.5 h.	1:1 to 3:1	[8]

2. ESTERIFICATION REACTION

As previously shown in Chapter 2 the esterification reaction has the following expression.

$$\underset{\text{Fatty acid}}{R\text{-}COOH} + \underset{\text{Alcohol}}{R_2OH} \overset{\text{Catalyst}}{\leftrightarrow} \underset{\text{Water}}{H_2O} + \underset{\text{Ester}}{R\text{-}COO\text{-}R_2} \quad (1)$$

Where the interaction of the fatty acid is taking place due to the presence of a heterogeneous catalyst. How the reaction takes place could be quite complicated depending on the type of catalyst and its functionality [5]. Nevertheless, it is important to point out that both types, basic and acid solid catalyst, are suitable to carry on the reaction.

Table 1, extracted part from reference [2] shows a great variety of reactions for several raw materials (fatty acid) different alcohols, as well as different operational conditions.

Table 1 shows a lot of work carried on with several different fatty acids as raw materials and in the presence of several types of alcohols as well as catalyst and different operational conditions. Many of them agree on the effect on the temperature over the reaction, producing a net increase in the final conversion when using higher temperatures [6,8-10,12,16,20,22-25]. A similar effect was also detected when the amount of catalyst was increased, in this case the reaction rate was modified and not the final conversion achieved. A higher amount of catalyst produces a net increase over the reaction rate and therefore shorter the reaction times [6,12,16,19,21-24,].

Among the previous works several catalysts have been tested: resin Amberlyst [6-8,16,19,21-23] has been the most employed one. However, other catalysts have also been used [7]. Due to the fact that resin Amberlyst is the most used one, in Table 2 it is compared three cases where this catalyst has been employed.

Kirbaşlar et al. [28] as well as Pöpken et al. [16] and Özbay et al. [29] have studied the esterification reaction using Amberlyst 15 as catalyst. The studied temperature range varies from 303 to 353 K for all the experiments done by the authors.

Table 2. Comparison of raw materials

	Özbay et al [29]	Pöpken et al. [16]	Kirbaşlar et al. [28]
Fatty acid used	*n.d.*	*Acetic acid*	*Acetic acid*
Alcohol	*Methanol*	*Methanol*	*Ethanol*
Temperature [K]	*333*	*323*	*353*
Catalyst	*Amberlyst-15*	*Amberlyst-15*	*Amberlyst-15*
Reaction time [h]	*3*	*6-7*	*6*
Amount of catalyst	*2%*	*7.71 g*	*5.4 g*
Conversion reached	*45.7%*	*>95%*	*75*

n.d: not disclosed, the type of fatty acid was not reported because a waste sample of oil from university cafeteria was used and it was not established the type of fatty acid in the mixture.

Table 3. Results

Catalyst	Alcohol	Reaction time [min]	Final Conversion
Zeolites NaY	Ethanol	180	24%
Zeolites USY	Ethanol	180	21%
Monosphere 550 A	Ethanol	240	15%
Monosphere 550 A	Ethanol 96	180	4%
Monosphere 550 A	1 propanol	240	5.6%
Monosphere 550 A	2 propanol	240	4.7%
Monosphere 550 A	Butanol	180	11.3%

Table 2 points out a comparison of the best scenario for each case showing the type of fatty acid as well as alcohol used for each case. A comparison of the final conversion achieved is also presented. It can be seen a very different final conversion due to several factors, reaction time for example, but more on the preparation of the catalyst as well as the system to carry on the experiments. Pöpken et al. [16] have used a reactive distillation system while Kirbaşlar et al. [28] used a batch reactor. This show the importance of the type of system to be used in order to improve the final conversion reached.

It is important to see that most of the work presented in Table 1 use pure fatty acid as catalyst; however, when analyzing the biodiesel production from an impure raw material, like an acid oil, the esterification and transesterification reaction takes place simultaneously, and therefore the presence of triglycerides becomes important to the reaction medium, either to contribute to the total volume of the sample as well as to produce some biodiesel from the transesterification reaction. Marchetti and Errazu [30] have done a comparison of several heterogeneous catalysts for the esterification reaction of Oleic acid in the presence of ethanol. A summary, extracted from Marchetti [2] is presented in Table 3.

In Table 3 it can be seen not very high conversion for the catalyst tested, this could be due to several issues as deactivation of the catalyst, the steric hindrance of the OH group in the alcohol, as well as that the operational conditions selected were not the best option. Although this not optimum results, it is encourage to continue the research in these options since they should be discharged as possible candidates.

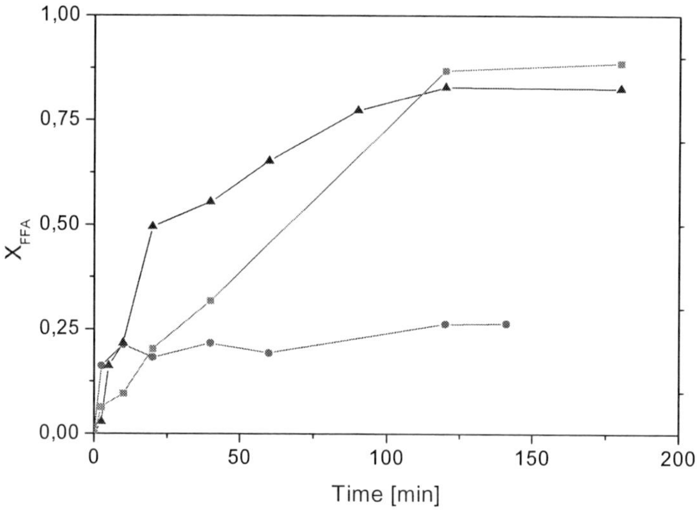

Figure 4. Conversion of FFA when using a solid heterogeneous catalyst. (●) T = 30, (▲) T = 45 and (■) T = 55.

Considering a sample with triglycerides as well as fatty acid, it is important to see the influence of the triglycerides in the reaction medium, as mentioned before. Marchetti et al. [31] have carried on the esterification reaction using an acid oil with 10% wt. of FFA as initial impurity. To analyze the esterification reaction alone, a titration procedure was done in order to establish the consumption of fatty acid. It was found that the final conversion, under similar conditions as showed when pure oleic acid was used, was improved, reaching a conversion five times higher than when no triglycerides are presented. Several factors might explain this tendency, being the more important one that the reaction takes place in the organic medium and the presence of triglycerides increases the organic volume section allowing more catalyst to be more time in contact with the oleic acid

Figure 4 shows the variations of the fatty acid conversion over time for three different temperatures showing the endothermic behavior of the reaction. This Figure was extracted from reference [31].

3. Transesterification Reaction

The transesterification reaction also takes place in the presence of a heterogeneous catalyst, being this a solid resin (basic or acid) as well as a zeolites or a new option such as a monolithic catalyst.

When using a solid catalyst it is important to know how the surface of the catalyst is. If the surface is a nonporous surface, then the reaction will always be in the surface of the catalyst.

Several works have been done regarding the transesterification reaction using solid catalyst. Among the work presented in the literature, some of the will be introduced here with the mayor variables used and the yield obtained with the aim of producing a summary of this reaction.

Samart et al. [32] studied the transesterification reaction of soybean oil and methanol using a KI/mesoporous silica catalyst. Several temperatures as well as reaction time and amount of catalyst have been tested reaching good results. The best conditions correspond to 10 h of reaction with 15% of catalyst and a molar ratio of 16:1 and temperature of 70°C, reaching a conversion of 94%.

A similar work, but with sunflower oil was done by Lukić et al. [33]. The authors used an alumina/silica catalyst over K_2CO_3 to carry on the transesterification reaction in the presence of methanol. Several temperatures (range from 80 to 200 °C) have been tested as well as several molar ratios: 6:1 to 30:1. One new variable used in this work was the calcinations temperature, showing that the best yield was obtained for a value of 600°C. The authors reached a conversion over 95% when 15:1 and 30:1 molar ratio with a reaction time of over 8 hours.

Kouzu et al. [34] have studied the transesterification reaction of rapeseed oil with methanol using crushed lime stone. They achieved a 96.5% conversion in 2h of reaction and were able to repeat this for over 15 times, showing a good stability of the catalyst. In their work, Velijković et al. [35] studied the same reaction as the previous authors but using calcium oxide as catalyst. The authors found that the reaction has two control regimes, starting by a mass transfer control followed by a kinetic control. Because of this the authors have used the sigmoid kinetic, reaching good agreement between data and model. In their data the authors reach almost full conversion.

Xu et al. [36] studied the reaction of soybean oil and 20% of mystiric acid with methanol in the presence of several heterogeneous catalyst based in Ta_2O_5. They have reached conversions as high as 80% when having 20% of FFA in the soybean oil, and carrying the reaction of esterification for 30 min and the transesterification for 24 h. the reaction temperature was 65°C, a molar ratio of 90:1 and 2% of catalyst was used.

Kolaczkowski et al. [37] studied the use of a new heterogeneous ZnL_2 catalyst to carry on the transesterification reaction. They achieved good results, as showed in Figure 5 extracted from reference [37] where a molar ratio of 12:1 was used and an initial pressure of 20 bar and temperature of 195°C.

Zabeti et al. [38] have done a review of several solid catalysts for biodiesel production; in their work they did a recompilation of several works where different substrates as well as different raw materials are used to produce biodiesel in the presence of an alcohol. The authors have summarized work performed with soybean, sunflower, palm kernel, colza and cottonseed oil (among other) and in the presence of metal oxides such as CaO, SrO, $Al_2O_3/TiO_2/ZnO$, among several other. The yield obtained is in all cases higher than 80% showing that all the alternatives presented there are suitable for biodiesel production.

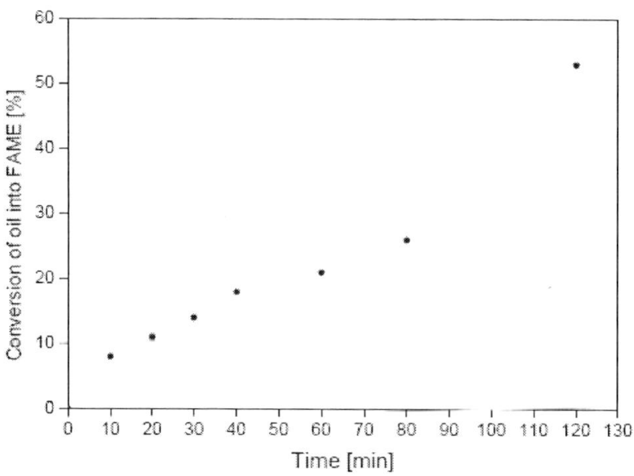

Figure 5. Evolution of the conversion into FAME vs. time. Ref. [37].

Table 4. Previous research on transesterification with solid catalysts

Vegetable Oil	Alcohol	Initial FFA	% Catalyst	Temp. [ºC]	Conversion	Reaction time	Molar ratio	Ref
Mix of oil	Methanol	0	Resin	--	98.3	--	--	4
Ethyl acetate	Methanol	0	MaO	20-50	90-98%	70 h	10:1 to 30:1	44
Canola	Methanol	10-20%	1-5% w/w TPA/HZ TPA/Si TPA/Al PA/AC	150-225	40-95%	600 min.	6:1 to 15:1	45
Ethyl butyrate	Methanol	0	0.4-1.6 wt of Calcium over: SBA15, MAM.41 Fummed silica's	60	65.7-95	1-5 h	2:1 to 12:1	46
Soybean Oil	Methanol	0	La/Zeolites beta	160	48.9	4 h	14.5:1	39
Soybean Oil	Methanol	0	$Mg.MgAl_2O_4$	65	57	10 h	3:1	39
Soybean Oil	Methanol	0	MgO, ZnO Al_2O_3	70 100 130	82	7 h	55:1	39
Soybean Oil	Methanol	0	ETS-10	120	94.6	24 h	6:1	39
Soybean Oil	Methanol	0	CaO SrO	65	95	0.5-3 h	12:1	39
Sunflower Oil	Methanol	0	CaO/SBA-14	160	95	5 h	12:1	39
Cottonseed	Methanol	0	1.5 wt% Base resin	30-70	20-95%	3 h	3:1 to 12:1	47
Soybean	Methanol	54.9	Amberlyst-15	80	98%	9 h	3:1 to 9:1	48
Soybean	Methanol	0	PbO MgO MnO_2 BaO CaO	215	84 66 80 85 78	14 min		49

Vegetable Oil	Alcohol	Initial FFA	% Catalyst	Temp. [ºC]	Conversion	Reaction time	Molar ratio	Ref
Soybean	Methanol	0	0.5-1.5% $Na/NaOH/\gamma-Al_2O_3$	60	80-95%	2 h	9:1	50
Canola	Ethanol	0	Mixed oxides of Mg-Co-Al-La	140-200	85-97%	5 h	7:1 to 16:1	51
Palm Kernel Oil	Methanol	0	5 to 20 wt% $LiNO_3/Al_2O_3$ $NaNO_3/Al_2O_3$ KNO_3/Al_2O_3	60	4.1-95%	1-3 h	30:1 to 80:1	52
Palm kernel	Methanol Ethanol	0	3wt% Al_2O_3 KF/Al_2O_3 KI/Al_2O_3 K_2CO_3/Al_2O_3 KNO_3/Al_2O_3	25 - 60	- 99.6 90 95.9 70	8h	6:1 to 15:1	53
Canola	Methanol	0	$CaTiO_3$ $CaMnO_3$ $Ca_2Fe_2O_5$ $CaZrO_3$ $CaCeO_3$	60	79 92 92 88 89	10 h	6:1	54
Coconut	Methanol	2.25%	3% w/w ZrO_2 ZnO SO_4^{-2}/SnO_2 SO_4^{-2}/ZrO_2 KNO_3/KL zeolites KNO_3/ZrO_2	200	49.3% 77.5% 80.6% 86.3% 77.2% 65.5%	4h	6:1	55

Table 4. Previous research on transesterification with solid catalysts (Continued)

Vegetable Oil	Alcohol	Initial FFA	% Catalyst	Temp. [ºC]	Conversion	Reaction time	Molar ratio	Ref
soybean	Methanol	2.6%	10 % de			24 h	6:1	56
			NaX	60-150 120-	6.8-23.9%			
			NaY	150 60-150	9.6-11.1%			
			KX	60-150 60-	10.3-31.5%			
			CsX	150 60-150	7.3-24.2%			
			(Cs,K)X	120-150 60-	8.6-27.3%			
			ETS-10	150 60-150	80.7-95.8%			
			K-ETS10	120-150 60-	93.5-93.8%			
			Cs-ETS10	150 120-	71.9-88.5%			
			(Cs,K)-ETS10	150 120-	67.4-88.1%			
			1 (NaOx/NaX)	150 120-	72.4-79.1%			
			3 (NaOx/NaX)	150 60-150	84.2-94%			
			4 (NaOx/NaX)		94.2-96.5%			
			0.25 (NaOx/NaX)		45.5-58.4%			
			1 (NaOx/NaX)		57.3-70.3%			
			3 (NaOx/NaX)		84.2-95.6%			
Soybean	Methanol	0	5% w/w de	180-200		1h	0.93:1	57
			CHT		92-96%			
			MgO		70-90%			
			MgO(I)		78-95%			
			MgO(II)		80-82%			
			MgO(III)		90-98%			
Soybean	Methanol	0	CaO	--	90-95%	2.5 h	4,5:1	58
			Ca(OCH3)2					
			Ba(OH)2					
			NaOH					

Table 5. Experimental conditions for the esterification reaction in the presence of triglycerides. Extracted from [60]

Experiment	Temperature [°C]	Molar ratio	% Catalyst	% of Oleic acid
1	45	6.12:1	2.26	10 wt%
2	45	6.12:1	5.1	10 wt%
3	45	6.12:1	7.05	10 wt%
4	45	4.12	2.26	10 wt%
5	45	5	2.26	10 wt%
6	30	6.12:1	2.26	10 wt%
7	55	6.12:1	2.26	10 wt%
8	45	6.12:1	2.26	2.81 wt%
9	45	6.12:1	2.26	27.22 wt%

Another review of solid catalyst for the transesterification reaction was done by Helwani et al. [39], the authors have made a recompilation of several works using similar raw materials to the previous work, but with some different catalyst and other operational conditions. In their work they showed that all the research they have studied show good yield towards the product, however, difference in the reaction time, temperature or molar ratio are to be notice.

Carbon based catalyst were prepared by Shu et al. [40] showing that they could be used for the transesterification reaction of cottonseed oil and methanol. Several operational conditions have been tested, and they have reached a final conversion of ~ 90% in 3h of reaction. This catalyst presented the difference to be an acid catalyst, showing that both, base and acid alternatives are suitable to produce biodiesel.

Similar work was done by Cao et al. [41], where both reaction where carried on using a super acid heterogeneous catalyst, heteropolyacid $H_3PW_{12}O_{40}\cdot 6H_20(PW_{12})$. The authors found a yield of 87% and 97% for the transesterification and direct esterification respectively, showing that this choice is suitable to carry on both reactions with good yields when the reaction temperature is 65°C. The reaction time was 16 hours while the molar ratio of methanol/oil was quite high, 70:1. Katada et al. [42] have also done some work in heteropolyacid catalyst, in their case the catalyst was $H_4PNbW_{11}O_{40}/WO_3\text{-}Nb_2O_5$. López et al [43], have studied both reaction simultaneously where both reaction were carried on using a zirconia modified catalyst. It was found that at 120°C and after 24 hours of reaction a complete conversion was achieved when a molar ratio of 12:1 was used. The reuse of the catalyst, one relevant point to consider when using heterogeneous catalyst, was not an easy step. The most active choice was the fresh zirconia.

Many more work have been done in heterogeneous catalysis, a summary of some of them are in the following Table 4.

Most of the previous work is related to the use of a pure vegetable oil as source for biodiesel production, presenting the mentioned disadvantage of the high price. Marchetti [2] and Marchetti et al. [59] studied the production of biodiesel using an acid oil and a solid base resin. The transesterification reaction was carried on simultaneously with the esterification reaction using ethanol as alcohol due to the possibility of producing it from natural source as sugar cane.

Experiments were carried on at different temperatures as well as several alcohol/oil molar ratios, amounts of catalyst as well as different catalyst. It has been found that the heterogeneous catalyst employed, was suitable to carry on the esterification reaction of oleic acid in the presence of triglycerides. Table 5 shows the variations of experiments performed.

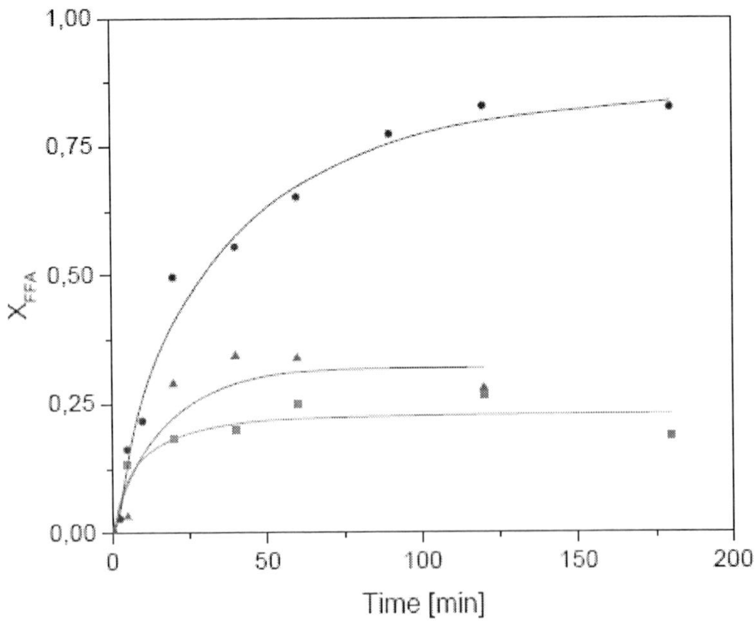

Figure 6. Conversion of fatty acid in the presence of triglycerides when using a heterogeneous catalyst. (■) N = 4.12:1, (▲) N = 5:1, (•) N = 6.21:1.

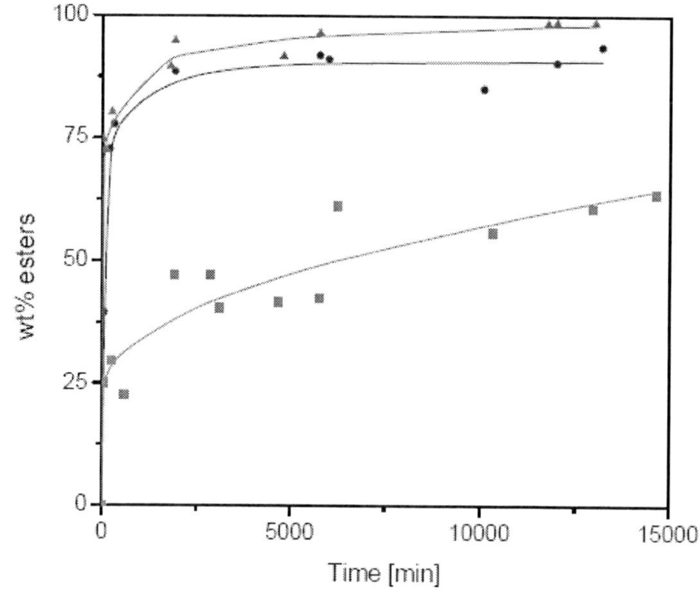

Figure 7. Esters amount vs. time when having different initial amount of FFA in the mixture. (■) $FFA^0 = 3.83\%$, (●)$FFA^0 = 10.68\%$ y (▲)$FFA^0 = 27.22\%$.

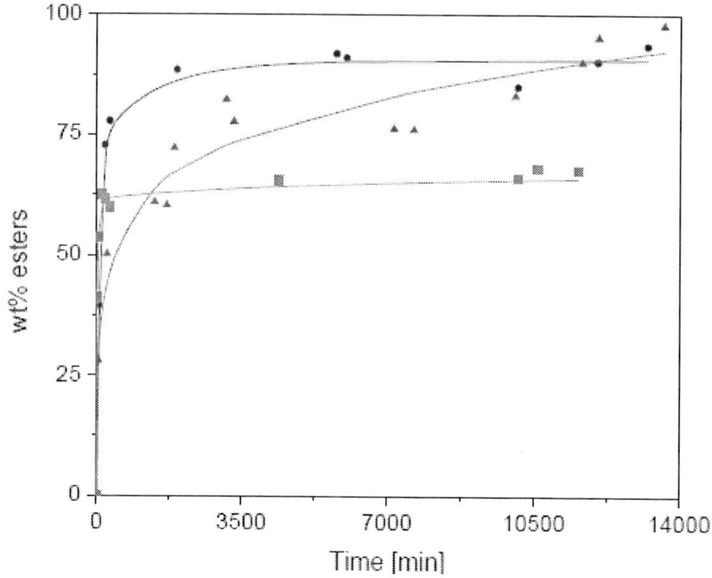

Figure 8. Esters amount vs. time for different reaction temperatures. (■) T = 30, (●) T = 45 y (▲) T = 55°C.

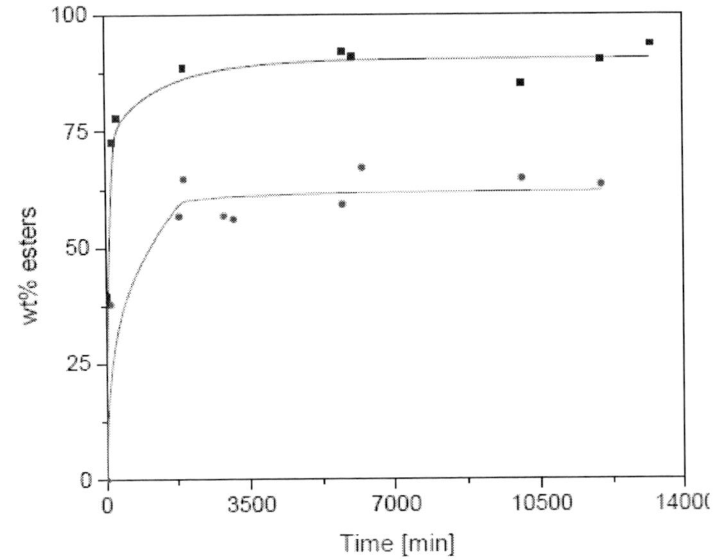

Figure 9. Esters amount vs. time for amount of catalyst. (●) S = 0.99%, (■) S = 2.12 %w/w.

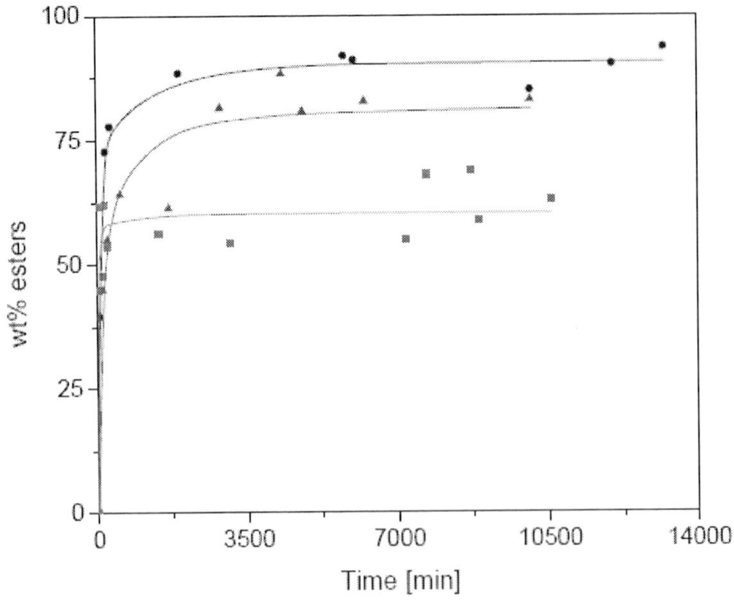

Figure 10. Esters amount vs. time for different amount of alcohol. (■) N = 4.12:1, (▲) N = 5.01:1, (●) N = 6.13:1.

From the experimental conditions from Table 5. Figure 6, was produced. In this figure it is presented the variations in the conversion of free fatty acids due to changes in the amount of alcohol (as illustrative from the data in Table 5). It can be seen that as the amount of alcohol increases, the final conversion increase as well, reaching a higher final value, see Figure extracted from reference [31].

Simultaneously to the esterification action, the transesterification of the sunflower oil was also followed and establish the production of FAME from both reaction as well as the consumption of triglycerides.

The same variables have been studied and the results obtained are presented in Figures 7 to 10 extracted from reference [61]

From Figure 7 to Figure 10 can be seen that when temperature increases as well as the amount of alcohol and the amount of catalyst the final percentage of ester produced from both reaction increases as well, reaching conversion over 99% in some cases. However, the mayor disadvantage from this catalyst is the reaction time, as it could be seen it is quite big and therefore it is not suitable from an industrial point of view since the time should be as short as possible. An optimization of this procedure should be done in order to reduce the reaction time to at least two hours that is the reaction time of the homogeneous catalyst [2]. In depth studies should be made in order to reduce the reaction time. Even more, other catalyst should be tested and tried in order to find new alternatives for this reaction with the aim of having a suitable alternative for biodiesel production.

Despite the technical aspects previously showed, it is important to establish some mathematical expression in order to be able to model a reactor and to have a mathematical tool to understand the behavior of the reactions taking place in the system. A kinetic section will be presented with the aim of providing a review of the kinetic modeling expression that could be found in the literature.

4. Kinetics

Kinetics expressions are quite relevant for modeling and for prediction. For this matter several works have been done with the aim of obtained an expression suitable to reproduce experimental data and to be used in modeling software. Because both reactions are relevant and important for biodiesel production plant, they will be presented separately.

4.1. Esterification Kinetics

Several works could be found in the literature for esterification kinetics, some of them will be presented here. The main purpose of a kinetic model is to establish the kinetic parameters such as activation energy and reaction rate constant, how the interaction between the reactant species takes places and how the catalyst participates in the reaction. Taking in consideration all the previous point, the kinetic expression could have several mathematical expression, first order kinetic; second order, reversible, pseudo first order, etc.

Tesser et al. [62] proposed a second order reversible reaction. However, due to the assumptions proposed by the authors, the final kinetic expression also considered adsorption over the catalyst. There are 3 mayor assumptions in this work, as follows: *i)* all compounds will be adsorbed or desorbed from the resin, *ii)* The adsorption follows a Langmuir isotherm, *iii)* the rate determining step is the reactive event between adsorbed oleic acid and methanol coming from the bulk phase. The final expression is as follows:

$$r_{cat} = \frac{k_{cat} b_A C_A C_M - k_{-cat} C_E b_E C_W}{1 + b_A C_A + b_M C_M + b_E C_E + b_W C_W} \tag{2}$$

In that work, the esterification between fatty acid and methanol in the presence of Amberlyst-15 was carried on. The results presented show a good agreement between the model and the experimental data.

On their work, Teo and Saha [12] proposed a kinetic model for the esterification of acetic acid and isoamyl alcohol when using a Purolite CT-175 as catalyst. A Langmuir-Hinshelwood-Hougen-Watson (LHHW) model was used. The result shows good agreement with the experimental data. As in the previous work, the controlling step is the reaction and not the adsorption or desorption of reactants or products.

Pöpken et al. [16] have studied one of the most common catalyst, Amberlyst-15, in the esterification of acetic acid and methanol. In their work, a secondary reversible model considering non ideality in the liquid phase as well as adsorption and desorption of the reactants and products provided an expression that fits the experimental data nicely. The final expression, writing in terms of activities is as follows:

$$r_{cat} = \frac{k_{cat} b_A C_A C_M - k_{-cat} C_E b_E C_W}{1 + b_A C_A + b_M C_M + b_E C_E + b_W C_W} \quad (3)$$

Many other works could be found in the literature, such as the one done by Kirbaşlar et al. [28], Omota et al. [63], Mengyu et al. [64], Ali [65], among other. In those works, several raw materials, different fatty acid as well as alcohols and catalyst have been employed in the experimental sections.

A comparison of the type of model, the value of the advance constant rate and the value of the activation energy is presented in Table 6, as long as with the fatty acid and alcohol employed.

The summary of kinetic expression is related to the use of fatty acid as raw material without the presence of other compounds. The presence of triglycerides is of high relevance and its effects are important due to the fact that in an acid oil the amount of fatty acid is quite high but the mayor compound continues to be the vegetable oil.

Table 6. Comparison of kinetics models

Fatty acid	Alcohol	Model	Reaction rate	Ea [kJ·mol-1]	Ref
Acetic acid	Methanol	Adsorption and kinetic with non ideality	8.497×10^6 [$mol \cdot g^{-1} \cdot s^{-1}$]	60.47	16
Acetic acid	Isoamyl	LHHW	2.4×10^5 [$mol \cdot g^{-1} \cdot min^{-1}$]	47.0	12
Oleic acid	Methanol	Eley-Rideal	131.6 [$cm^3 \cdot g^{-1} \cdot min^{-1}$]	74.65	62
Mix of FFA	Methanol	LH	7.847×10^4 [$L \cdot mol^{-1} \cdot h^{-1}$]	18.59	64
Dodecanoic	2-ethyl-hexanol	Non ideal second order reversible	1.643×10^5 [$m^6 \cdot kmol^{-1} \cdot kg^{-1} \cdot s^{-1}$]	65.5	63
Propionic*	Methanol	Used Pöpken model [16]	8.0×10^{-5} (at 35°C) [$mol^1 \cdot g^{-1} \cdot s^{-1}$]	34.7	65
Acetic	Ethanol		2.6×10^{14} [$(m^3)^2 \cdot kmol^{-2} \cdot s^{-1}$]	104.129	28

* Several mechanism has been tested, but the one presented is the best option for the experiments presented there.

Table 7. Kinetic parameters

Parameter	Value	Units
k_∞	2.22 10^{-2}	l/(kg hr)
E	1.00 10^{3}	cal/mol
K_∞	3.07 10^{66}	
ΔH	9.56 10^{4}	cal/mol
K_1	6.16 10^{-1}	l^2/mol^2
K_2	1.71 10^{2}	l/mol

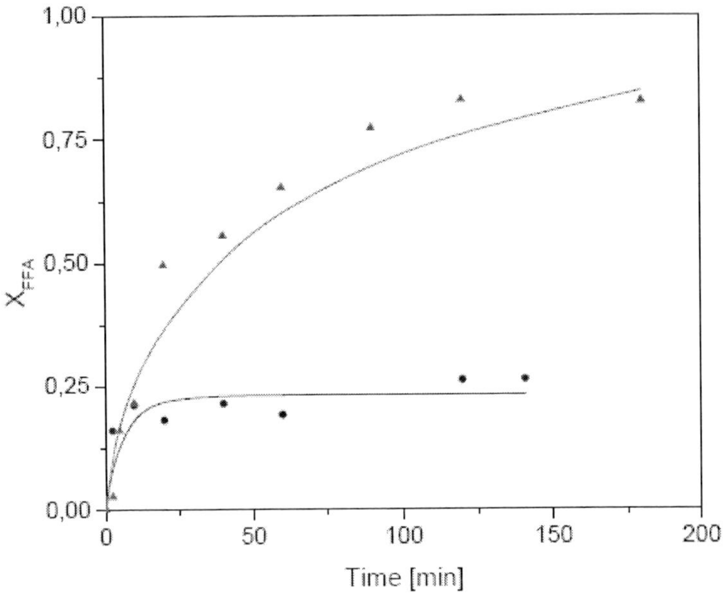

Figure 11. Conversion of FFA vs. time for different reaction temperatures (●) T = 30 and (▲) T = 45.

Because of this, Marchetti [2] has studied the kinetics of the esterification reaction when the triglycerides are presented. It has been found two possible alternatives, one when the triglycerides act as a inert to the system and the second one, when triglycerides are also converted into biodiesel.

For that work, Marchetti [2] proposed several mechanism considering all the possible interactions between the compounds and the catalyst, this has lead to more than 30 possible kinetic expression that have been tested. The best approximation achieved, for two reaction temperatures could be seen in Figure 11. This was obtained by using the following mechanism:

ALC + s → ALCs (4)

FFAL + ALCs → FAEE + H_2Os (5)

H_2Os → H_2O + s (6)

where both polar compounds are being adsorbed or desorbed from the catalyst, this leads to the following kinetics expressions.

$$r_{FFAl} = m_{cat} k \left(C_{alc} - \frac{C_{faee} C_w}{C_{FFA} K} \right) \quad (7)$$

$$m_{cat} = \frac{m^0_{cat}}{1 + K_1 C_w + \dfrac{K_2 C_{faee} C_w}{C_{FFA}}} \quad (8)$$

with $k = k_\infty e^{\left(-\frac{E}{RT}\right)}$ and $K = K_\infty e^{\left(-\frac{\Delta H}{RT}\right)}$

Fitting the experimental data it has been obtained the following kinetics parameters.

As expected it could be seen that a higher temperature gives a higher final conversion. Similar results have been seen when modifying the amount of catalyst. However these results are in good agreements with previous results from Tesser et al. [62] showing similar behavior. In either of them it has been considered that the triglycerides will actually be converted into biodiesel, and therefore a more complete kinetic model is required.

4.2. Transesterification Kinetics

Different works have been reported for the kinetic of the transesterification reaction using solid catalyst. The most common mechanism is related to consider the reaction as elementary step from TG to DG without any intermediate step. Bokade et al. [66] presented a simple model for the transesterification of edible and non edible oils. They have proposed a simple reversible mechanism; they have found good agreement with the experimental data. On the other hand, it could be found much simpler kinetics proposals, such as considering the whole global reaction as the main and only reaction taking place. This is the case of Huang et al. [67] works, in their research; a three region kinetics was established due to the fact that several different types of control could be taking place, mass transfer, reaction, etc. But in order to understand this regimes, a simpler kinetic model was proposed. Finally, it could be also obtained a kinetic expression based on heterogeneous model, such as Eley-Rideal type of adsorptions. Dossin et al. [44] have proposed this kind of mechanism in order to fit their experimental data and manage to do it quite well. They have assumed reversible reactions and the methanol to be adsorbed on the catalyst.

Because of this, and as done in section 4.2, a more general mechanism is required when both reactions take place simultaneously. And since the alcohol is adsorbed in the solid resin a new kinetic mechanism is required. It has been proposed the following alternative:

$$\mathrm{ALC + s \leftrightarrow ALCs} \tag{9}$$

$$\mathrm{TG + ALCs \leftrightarrow DG + FAEE + s} \tag{10}$$

$$\mathrm{DG + ALCs \leftrightarrow MG + FAEE + s} \tag{11}$$

$$\mathrm{MG + ALCs \leftrightarrow G + FAEE + s} \tag{12}$$

$$\mathrm{FFA + ALCs \leftrightarrow FAEE + H_2O + s} \tag{13}$$

Assuming that the controlling step is the surface reaction, then the final expressions are as follows:

$$r_1 = \left(k_1 C_{tg} C_{alc} K_1 - k_2 C_{dg} C_{faee}\right) \frac{m_{cat}^0}{\left(1 + K_1 C_{alc}\right)} \tag{14}$$

$$r_2 = \left(k_3 C_{dg} C_{alc} K_1 - k_4 C_{mg} C_{faee}\right) \frac{m_{ca}^0}{\left(1 + K_1 C_{alc}\right)} \tag{15}$$

$$r_3 = \left(k_5 C_{mg} C_{alc} K_1 - k_6 C_g C_{faee}\right) \frac{m_{cat}^0}{\left(1 + K_1 C_{alc}\right)} \tag{16}$$

$$r_4 = \left(k_7 C_{FFA} C_{alc} K_1 - k_8 C_w C_{faee}\right) \frac{m_{cat}^0}{\left(1 + K_1 C_{alc}\right)} \tag{17}$$

$$m_{cat} = \frac{m_{cat}^0}{\left(1 + K_1 C_{alc}\right)} \tag{18}$$

Fitting the experimental data, presented in Figure 12 to 14, it could be obtained the following kinetics parameters

Table 8. Kinetics parameters

Parameter	Value	Units
$k_{\infty 1}$	1.95 10^{-2}	l/(kg h)
E_1	1.49 10^{3}	cal/mol
$k_{\infty 2}$	9.03 10^{1}	l/(kg h)
E_2	2.03 10^{3}	cal/mol
$k_{\infty 3}$	1.49 10^{4}	l/(kg h)
E_3	9.85 10^{3}	cal/mol
$k_{\infty 4}$	8.74 10^{1}	l/(kg h)
E_4	1.95 10^{3}	cal/mol
$k_{\infty 5}$	1.01 10^{2}	l/(kg h)
E_5	9.20 10^{3}	cal/mol
$k_{\infty 6}$	3.83 10^{3}	l/(kg h)
E_6	9.41 10^{3}	cal/mol
$k_{\infty 7}$	2.79 10^{13}	l/(kg h)
E_7	2.35 10^{4}	cal/mol
$k_{\infty 8}$	1.58 10^{1}	l/(kg h)
E_8	1.42 10^{3}	cal/mol
K_1	8.56 10^{3}	l/mol

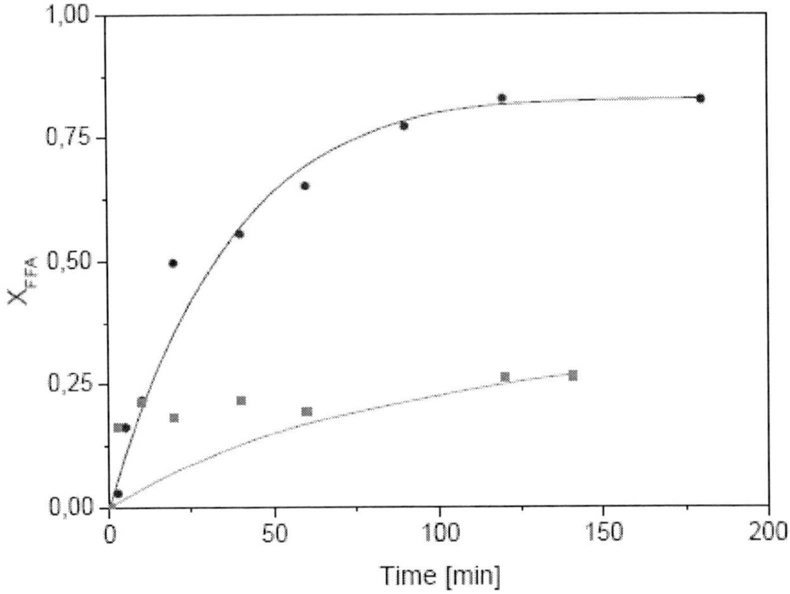

Figure 12. Conversion of FFA vs. time for different reaction temperatures. (■) T = 30 and (●) T = 45.

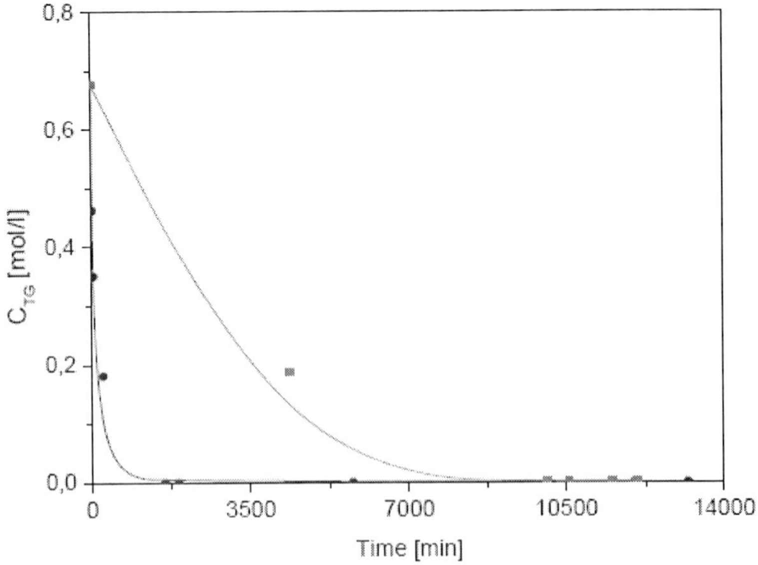

Figure 13. Variation of the triglycerides concentration for different reaction temperatures. (■) T = 30 and (●) T = 45.

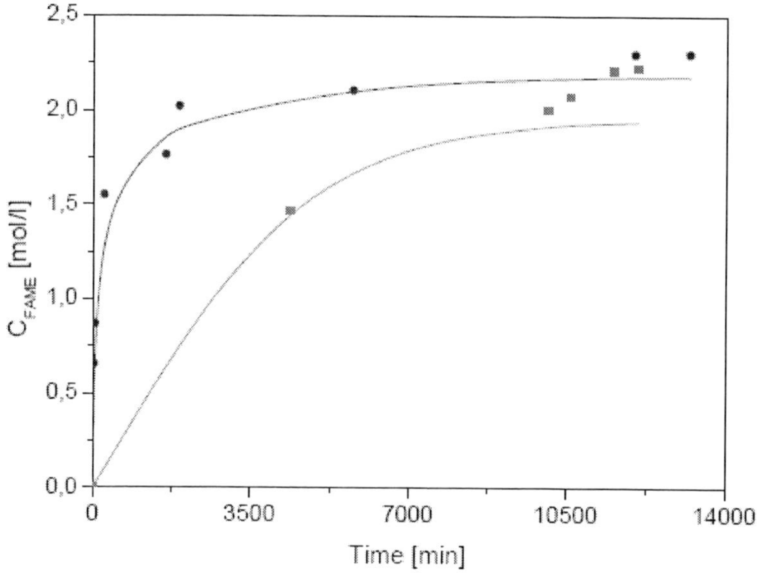

Figure 14. Variation of the concentration of biodiesel produced for different reaction temperatures. (■) T = 30 and (●) T = 45.

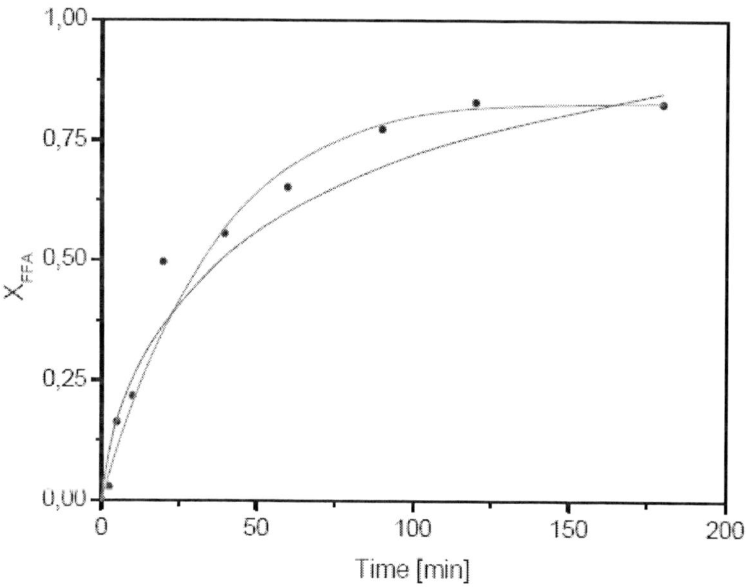

Figure 15. Comparison of models. (—) TG as inert, (—) TG being converted. (●) Experimental data.

When obtaining these results, it was interesting to compare the conversion of fatty acid when the triglycerides are consider inert and when they are being converted and modifying the reaction medium. In Figure 15 it could be seen this comparison. The mechanism involving the reaction of triglycerides shows a much better correlation of the experimental data. This was expected due to the fact that in the experimental setup there is a slight conversion of triglycerides. Even thought, more complex alternatives could be presented considering other possible side reactions.

5. Economic Study

As presented in several chapters, economic studies are relevant to analyze the commercial impact of the process.

Heterogeneous alternative appear as a promising technology, as showed in the work done by Marchetti et al. [60]. Showing that heterogeneous options are quite suitable and very promising regarding the treat of waste oil.

However, is important to notice that the most impure the process the more pre-purification that is required and that has not been considered in the previous mentioned work.

Even more, the assumptions considered as market situation, prices, availability of raw materials, selling market for the final fuel, etc. are quite important and have a net influence in the process, as presented by Marchetti [2] where it was showed that different variables have very important effect on the global economic of a biodiesel plant and are constantly changing according to global politics and decisions.

References

[1] Haas, M.J., McAloon, A.J., Yee, W.C., Foglia, T.A. *A process model to estimate biodiesel production costs. Bioresource Technology*. 97(4). (2006), 671-678.

[2] Marchetti, J.M. Ph.D. thesis. *Technological alternatives for biodiesel production*. Universidad Nacional del Sur. Argentina (2008).

[3] Marchetti, J.M., Miguel, V.U., Errazu, A.F. *Analysis of Technological Alternatives for Biodiesel Production.* 2nd Mercosur Congress on

Chemical Engineering- 4th Mercosur Congress on Process Systems Engineering. Río de Janeiro, Brasil. ISBN:85-7650-042-6. (2005).

[4] Bournay, L., Casanave, D., Delfort, B., Hillion, G., Chodorge, J.A. *New heterogeneous process for biodiesel production: A way to improve the quality and the value of the crude glycerin produced by biodiesel plants. Catalysis Today*. 105. (2005), 190-192.

[5] Chakrabarti, A., Sharma, M.M. *Cationic ion exchange resins as catalyst. Reactive Polymers*. 20. (1993), 1-45.

[6] Gangadwala, J., Mankar, S., Mahajani, S. Kienle, A., Stein, E. *Esterification of acetic acid with Butanol in the presence of ion-exchange resins as catalyst. Industrial and Engineering Chemistry Research.* 42. (2006), 2143-2155.

[7] Peters, T.A., Benes, N.E., Holmen, A., Keurentjes, J.T.F. *Comparison of commercial solid acid catalysts for the esterification of acetic acid with butanol* Applied catalysis A. General. 297. (2006), 182-188.

[8] Liu, W.T., Tan, C.S. *Liquid-phase esterification of propionic acid with n-butanol. Industrial and Engineering Chemistry Research.* 40. (2001), 3281-3286.

[9] Lee, M.J., Wu, H.T., Lin, H.M. *Kinetic of catalytic esterification of acetic acid and amyl alcohol over dowex. Industrial and Engineering Chemistry Research.* 39. (2000), 4094-4099.

[10] Mazzotti, M., Neri, B., Gelosa, D., Kruglov, A., Morbidelli, M. *Kinetics of liquid-phase esterification catalyzed by acidic resins. Industrial and Engineering Chemistry Research.* 36. (1997), 3-10.

[11] Lilja, J., Yu Murzin, D., Salmi, T., Aumo, J., Mäki-arvela, P., Sundell, M. *Esterification of different acids over heterogeneous and homogenous catalysts and correlation with the Taft equation. Journal of Molecular Catalysis A: Chemical.* 182-183. (2002), 555-563.

[12] Teo, H.T.R. Saha, B. *Heterogeneous catalyzed esterification of acetic acid with isoamyl alcohol: kinetic studies. Journal of Catalysis*. 228. (2004), 174-182.

[13] Altiokka, M.R., Çitak, A. *Kinetics study of esterification of acetic acid with isobutanol in the presence of amberlite catalyst.* Applied Catalysis A. General. 239. (2003), 141-148.

[14] Liu, Y., Lotero, E., Goodwin Jr. J.G. *A comparison of the esterification of acetic acid with methanol using heterogeneous versus homogeneous acid catalysis. Journal of Catalysis.* 242. (2006), 278-286.

[15] López, D.E., Suwannakarn, K., Bruce, D.A., Goodwin Jr., J.G. *Esterification and transesterification on tungstated zirconia: effect of the calcination temperature*. *Journal of Catalysis*. 247. (2007), 43-50.

[16] Pöpken, T., Götze, L., Gmehling, J. *Reaction kinetics and chemical equilibrium of homogeneously and heterogeneously catalyzed acetic acid esterification with methanol and methyl acetate hydrolysis*. *Industrial and Engineering Chemistry Research*. 39. (2000), 2601-2611.

[17] Chen, X., Xu, Z., Okuhara, T. *Liquid phase esterification of acrylic acid with 1-butanol catalyzed by solid acid catalyst*. Applied Catalysis A: General. 180. (1999), 261-269.

[18] Benedict, D.J., Parulekar, S.J., Tsai, S.P. *Esterification of lactic acid and ethanol with/without pervaporation*. *Industrial and Engineering Chemistry Research*. 42. (2003), 2282-2291.

[19] Goto, S., Takeuchi, M., Matouq, M.H. *Kinetics of esterification of palmitic acid with isobutyl alcohol on ion-exchange resins pellets*. *International Journal of Chemical Kinetics*. 24. (1992), 587-592.

[20] Zhang, Y., Ma, L., Yang, J. *Kinetics of esterification of lactic acid with ethanol catalyzed by cation exchange resins*. *Reactive and Functional Polymers*. 61(1). (2004), 101-114.

[21] Delgado, P., Sanz, M.T., Beltrán, S. *Kinetic study for esterification of lactic acid with ethanol and hydrolysis of ethyl lactate using an ion-exchange resin catalyst*. *Chemical Engineering Journal*. 126. (2007), 111-118.

[22] Sanz, M.T., Murga, R., Beltrán, S. Cabezas, J.L. *Autocatalyzed and ion-exchange resins catalyzed esterification kinetics of lactic acid with methanol*. *Industrial and Engineering Chemistry Research*. 41. (2002), 512-517.

[23] Yadav. G.D., Thathagar, M.B. *Esterification of maleic acid with ethanol over cation-exchange resins catalysts*. *Reactive and Functional Polymers*. 52. (2002), 99-110.

[24] Levesque, C.L. Craig, A.M. *Kinetics of an esterification with cation-exchange resin catalyst*. *Industrial and Engineering Chemistry*. 40(1). (1948), 96-99.

[25] Sánchez, N., Coteron, A., Martinez, M., Aracil, J. *Kinetic analysis and modeling of the esterification of oleic acid and aleyl alcohol using cobalt chloride as catalyst*. *Industrial and Engineering Chemistry Research*. 31. (1992), 1985-1988.

[26] Mbaraka, I.K., Radu, D.R., Lin, V.S.Y., Shanks, B.H. *Organosulfonic acid-functionalized mesoporous silicas for the esterification of fatty acid. Journal of Catalysis*. 219. (2003), 329-336.

[27] Caetano, C.S., Guerreiro, L., Fonseca, I.M., Ramos, A.M, Vital, J., Castanheiro, J.E. *Esterification of fatty acids to biodiesel over polymers with sulfonic acid groups.* Applied Catalysis A: General. 359. (2009), 41-46.

[28] Kirbaşlar, S.I., Baykal, Z.B., Dramur, U. *Esterification of acetic acid with ethanol catalyzed by an acidic ion exchange resin. Turkish Journal Engineering Environmental Science.* 25. (2001), 569-577.

[29] Özbay, N., Optar, N., Tapan, N.A. *Esterification of free fatty acids in waste cooking oils (WCO): Role of ion exchange resins*. Fuel. 87. (2008), 1789-1798.

[30] Marchetti, J.M., Errazu, F.E. *Comparison of different heterogeneous catalysts and different alcohols for the esterification reaction of oleic acid. Fuel*. 87. (2008), 3477-4480.

[31] Marchetti, J.M., Miguel, V.U., Errazu, A.F. *Heterogeneous esterification of oil with high amount of free fatty acids. Fuel.* 86. (2006), 906-910.

[32] Samart, C., Sreetongkittikul, P., Sookman, C. *Heterogeneous catalysis of transesterification of soybean oil using KI/mesoporous silica. Fuel Processing Technology*. 90. (2009), 922-925

[33] Lukić, I., Krstić, J., Javanović, D., Skala, D. *Alumina/silica supported K_2CO_3 as a catalyst for biodiesel synthesis from sunflower oil. Bioresource Technology*. 100. (2009), 4690-4696.

[34] Kouzu, M., Hidaka, J.S., Komichi, Y., Nakano, H., Yamamoto, M. *A process to transesterify vegetable oil with methanol in the presence of quick lime bit functioning as solid base catalyst. Fuel.* 88. (2009), 1983-1990.

[35] Velijković, V.B., Stamenković, O.S., Todorović, Z.B., Lazić, M.L., Skala, D.U. *Kinetics of sunflower oil methanolysis catalyzed by calcium oxide. Fuel.* 88. (2009), 1554-1562.

[36] Xu, L., Yang, X., Yu, X., Guo, Y. Maynurkader. *Preparation of mesoporous polyoxometalate-tantalum pentoxide composite catalyst for efficient esterification of fatty acid. Catalysis Communications*. 9. (2008), 1607-1611.

[37] Kolaczkowski, S.T., Asli, U.A., Davidson, M.G. *A new heterogeneous ZnL_2 catalyst on a structured support for biodiesel production. Catalysis Today.* 147(1). (2009), 220-224.

[38] Zabeti, M., Daud, W.M.A.W., Aroua, M.K. *Activity of solid catalysts for biodiesel production: a review. Fuel Processing Technology*. 90. (2009), 770-777.

[39] Helwani, Z., Othman, M.R., Aziz, N., Kim, J., Fernando, W.J.N. *Solid heterogeneous catalysts for transesterification of triglycerides with methanol: a review.* Applied Catalysis A. General. 363. (2009), 1-10.

[40] Shu, Q., Zhang, Q., Xu, G., Nawaz, Z., Wang, D., Wang, J. *Synthesis of biodiesel from cottonseed oil and methanol using a carbon-base solid acid catalyst. Fuel Processing Technology*. 90. (2009), 1002-1008.

[41] Cao, F., Chen, Y., Zhai, F., Li, J., Wang, J., Wang, X., Wang, S., Zhu, W. *Biodiesel production from high acid value waste frying oil. Biotechnology and Bioengineering*. 101(1). (2008), 93-100.

[42] Katada, N., Hatanaka, T., Ota, M., Yamada, K., Okumura, K., Niwa, M. *Biodiesel production using heteropoly-acid derived solid acid catalyst* $H_4PNbW_{11}O_{40}/WO_3\text{-}Nb_2O_5$. Applied Catalysis A: General. 363. (2009), 164-168.

[43] Lopéz, D.E., Goodwing, J.G., Bruce, D.A., Furuta, S. *Esterification and transesterification using modified-zirconia catalysts.* Applied Catalysis A: General. 339. (2008), 76-83

[44] Dossin, T.F., Reyniers, M.F., Marin, G.B. *Kinetics of heterogeneously MgO-catalyzed transesterification.* Applied Catalysis B: Environmental 61. (2006), 35-45.

[45] Kulkarni, M.G., Gopinath, R., Meher, L.C., Dalai, A.K. *Solid acid catalyzed biodiesel production by simultaneous esterification and transesterification. Green Chemistry*. 8. (2006), 1056-1062.

[46] Albuquerque, M.C.G., Urbistondo, I.J., González, J.S., Robles, J.M.M., Tost, R.M., Castellon, E.R., López, A.J., Azevedo, D.C.S.,Cavalcante, C.L., Torres, P.M. *CaO supported on mesoporous silicas as basic catalysts for transesterification reactions.* Applied Catalysis A: General. 334. (2008), 35-43.

[47] Georgogianni, K.G., Katsoulidis, A.K., Pomonis, P.J., Manos, G., Kontominas, M.G. *Transesterification of rapeseed oil for the production of biodiesel using homogeneous and heterogeneous catalysis*. Fuel Processing Technology. 90. (2009), 1016-1022.

[48] Jaya, N., Ethirajulu, K., Sundar, S., Elanchelian, C. *Kinetic parameters of heterogeneously catalyzed transesterification of cottonseed oil to biodiesel.* Proceedings of international conference on energy and environment. (2009), 485-488.

[49] Park, J.Y., Kim, D.K., Lee, J.S. *Esterification of free fatty acid using water tolerable amberlyst as a heterogeneous catalyst. Bioresource Technology*. 101. (2010), 62-65

[50] Singh, A.K., Fernando, S.D. *Reaction kinetics of soybean oil transesterification using heterogeneous metal oxide catalysts. Chemical Engineering Technology*. 30(12). (2007), 1716-1720.

[51] Kim, H.J., Kan, B.S., Kim, M.J., Park, Y.M., Kim, D.K., Lee, J.S., Lee, K.Y. *Transesterification of vegetable oil to biodiesel using heterogeneous base catalyst. Catalysis Today*. 93-95. (2004), 315-320.

[52] Li, E, Xu, Z.P., Rudolph, V. *MgCoAl-LDH derived heterogeneous catalysts for the ethanol transesterification of canola oil to biodiesel.* Applied Catalysis B: Environmental. 88. (2009), 42-49.

[53] Benjapornkulaphon, S., Ngamcharussruvichai, C., Bunyakiat, K. *Al2O3-supported alkali and alkali earth metal oxides for transesterification of palm kernel oil and coconut oil. Chemical Engineering Journal*. 145. (2009), 468-474.

[54] Boz, N., Kara, M. *Solid base catalyzed transesterification of canola oil. Chemical Engineering Communications*. 196. (2009), 80-92.

[55] Kawashima, A., Matsubara, K., Honda, K. *Development of heterogeneous base catalysts for biodiesel production. Bioresource Technology*. 99. (2008), 3439-3443.

[56] Jitputti, J., Kitiyanan, B., Rangsunvigit, P., Bunyakiat, K., Attanatho, L., Jenvanitpanjakul, P. *Transesterification of crude palm kernel oil and crude coconut oil by different solid catalysts. Chemical Engineering Journal.* 116(1). (2006), 61-66.

[57] Suppes, G.J., Dasari, M.A., Doskocil, E.J., Mankidy, P.J., Goff, M.J. *Transesterification of soybean oil with zeolite and metal catalysts.* Applied Catalysis A: General 257. (2004), 213-223.

[58] Di Serio, M. Ledda, M., Cozzolino, M., Minutillo, G., Tesser, R., Santacesaria, E. *Transesterification of Soybean Oil to Biodiesel Using Heterogeneous Basic Catalysts. Industrial and Engineering Chemistry Research.* 45. (2006) 3009-3014.

[59] Gryglewicz, S. *Rapeseed oil methyl esters preparation using heterogeneous catalysts. Bioresource Technology*. 70(3). (1999), 249-253.

[60] Marchetti, J.M., Miguel, V.U., Errazu, A.F. *Techno-economic study of different alternatives for biodiesel production. Fuel Processing Technology.* 89. (2008), 740–748.

[61] Marchetti, J.M., Errazu, A.F. *Biodiesel production from acid oils and ethanol using a solid basic resin as catalyst. Biomass and Bioenergy.* (article in press)

[62] Tesser, R., Casale, L., Verde, D., Di Serio, M., Santacesaria, E. *Kinetics of free fatty acids esterification: batch and loop reactor modeling. Chemical Engineering Journal.* (article in press)

[63] Omota, F., Dimian, A.C. Bliek, A. *Fatty acid esterification by reactive distillation: Part 2-kinetics based design for sulphated zirconium catalysts. Chemical Engineering Science*. 58. (2003), 3175-3185.

[64] Mengyu, G., Deng, P., Li, M., En, Yue., Jianbing, H. *The kinetics of the esterification of free fatty acids in waste cooking oil using* $Fe_2(SO_4)_3/C$ *catalyst. Catalysis, Kinetics and Reactors*. 17(1). (2009), 83-87.

[65] Ali, S.H. *Kinetics of catalytic esterification of propionic acid with different alcohols over amberlyst 15. International Journal of Chemical Kinetics.* 41. (2008), 432-448.

[66] Bokade, V.V., Yadav, G.D. *Transesterification of edible and non-edible vegetable oils with alcohols over heteropolyacids supported on acid treated clay. Industrial and Engineering Chemistry Research.* 48. (2009), 9408-9415.

[67] Huang, K., Xu, Q., Zhang, S., Ren, Z., Yan, Y. *Multi-step controlled kinetics of the transesterification of crude soybean oil with methanol by* $Mg(OCH_3)_2$*. Chemical Engineering Technology.* 32(10). (2009), 2595-1604.

Chapter 7

ENZYMATIC CATALYSIS

1. INTRODUCTION

Besides the use of heterogeneous solid catalyst, as described in the previous chapter, new technologies are also being studied. Among those, enzymes appear as a very promising option due to several advantages that they has over conventional catalyst [1-2]. Some of those advantages are: *i)* enzymes could be regenerated and, therefore, reused for several times. *ii)* When using enzymes, if a high concentration is utilized, the activity could be longer on a reaction environment. *iii)* Enzymes are, by nature, thermal stable (not on the operational temperature). *iv)* Lipases could be immobilized over different supports and this will prevent agglomeration as well as it will improve the reuse of the enzyme. *v)* Since it could be immobilized, then it could also be considered as a heterogeneous catalyst; therefore, when is used in a continuous process, it could be used as a fix bed catalyst improving the separation of the reactants and products.

Although this advantages, there are some inconvenient regarding the use of enzymes for biodiesel production, such as, enzymes get deactivated with high amounts of water, therefore the presence of free fatty acid is compromised; the biocatalyst is more expensive than the regular enzyme, etc.

All these advantages have produced a direct interest in the study of the enzymes as catalyst for biofuel. Enzymes have been widely used for bioethanol production [3-7]. This is because bioethanol is produced normally by fermentation of sugar from sugar cane for example. However, since there is net interest in using other raw materials, starch or yeast, the main raw material is lignocelluloses or hemicelluloses. When these compounds are being used, the production process is not only fermentation as described for ethanol from sugar. In those cases there are several steps involved, such as an acid pretreatment (this could also be with steam). [8-9], and then there is an enzymatic hydrolysis in order to modified the lignocelluloses into sugars so they can fermented into alcohol [4,10]. A schematic idea of the process is presented in Figure 1. Extracted from reference [10]

Regarding the use of enzymes for other liquid biofuel, in this case our priority is related to biodiesel, different enzymes, different raw materials as well as operational conditions have been tested by several authors with the aim of studying the optimums operational conditions to produce the biofuel. The flow diagram in Figure 2 shows the enzymatic process for biodiesel production, where it can be easily seen that there is a very few equipments involved in comparison with the homogeneous processes.

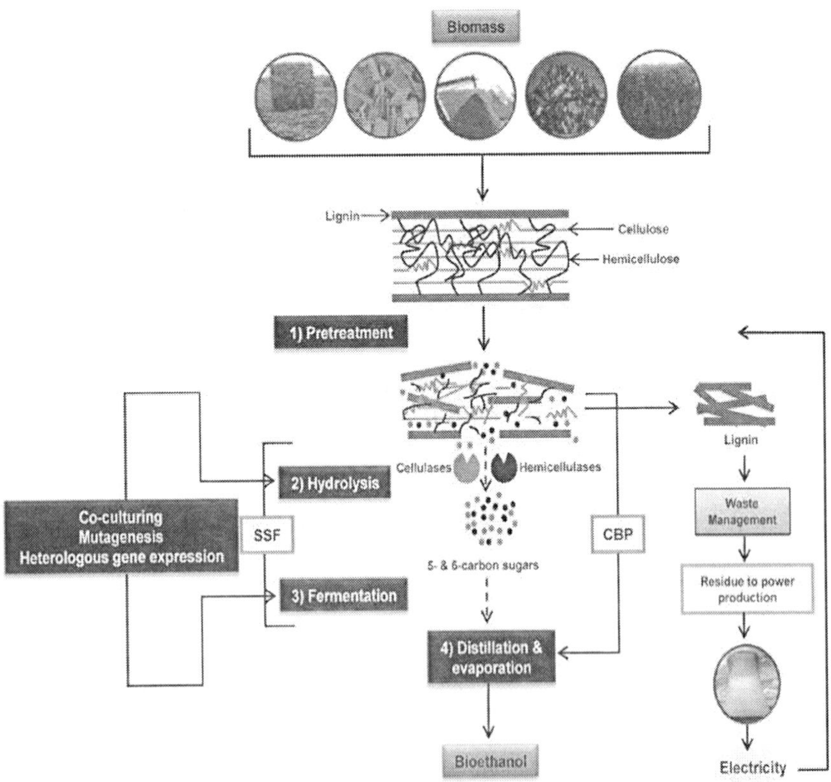

Figure 1. From Biomass to bioethanol, schematic representation.

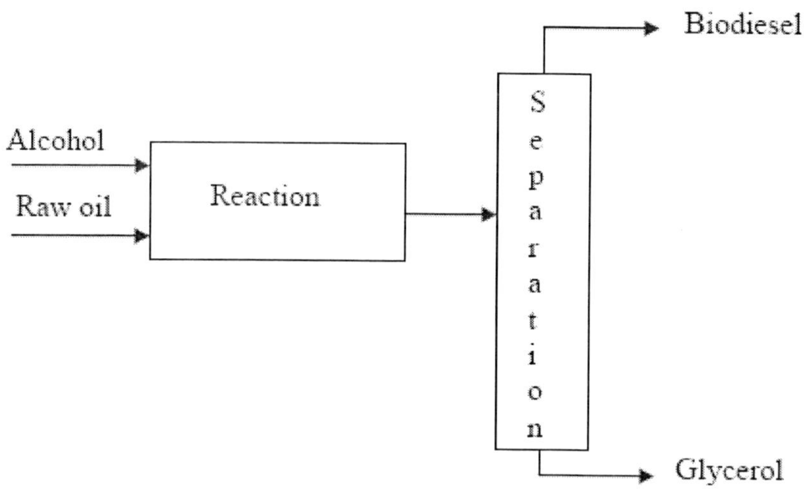

Figure 2. Schematic flow diagram of an enzymatic biodiesel process.

In Figure 2 there are very few equipments, and the main difference with the homogeneous process is that when purifications was needed for the homogeneous catalyst, several equipments are required, neutralization, decanters, distillation, separation, etc. On the other hand, when separation is required for a heterogeneous case, such as immobilized enzymes, the main separation is a distillation column, to recover the methanol that is in an excess amount and a decanter in order to separate the biodiesel from the glycerol. This makes the process easier, simpler, having a lower investment in equipment and more environmentally friendly.

In the schematic flow diagram of Figure 2, there is no pretreatment as presented in chapter 5 when using sulfuric acid; this is because, if the enzyme might be able to handle some water, then the esterification reaction will take place simultaneously with the transesterification reaction. We have divided the following sections, as it has been done previously, in the study of each reactions individually. However, when considering the reaction times for the esterification reaction and the transesterification when using enzymes, these times are more close to each other than for the other previous cases. Therefore, the need of overlapping both reactions is stronger than in the previous scenarios, but for a more simple explanation of the process we will discuss them separately.

2. ESTERIFICATION REACTION

As it was previously introduced, this reaction consists in the interaction of a fatty acid with an alcohol, in order to produce an ester (Fatty acid alkyl ester) and water. The alkyl ester is normally called methyl ester due to the fact that methanol is the most common alcohol employed.

Although the esterification reaction using enzymes is not a mayor contribution to the biodiesel industry in today's economy, it is important to make a few notes on it, since this technology might become the technology of tomorrow.

Along a lot of works regarding this topic, some of them will be summarize below in order to provide some trends on what it has been done so far in this matter.

One of the most common enzymes used for the esterification reaction is *Candida antartica* [11-14]. This enzyme has been used for several fatty acids in order to produce esters, several times in its commercial form as Novozym 435, showing a good final yield. Garcia et al. [11], used this commercial enzyme for the esterification of palmitic acid and iso-propyl alcohol. They have studied the influence of several operational variables, such as temperature range (65 to 75°C) molar ratio of alcohol/fatty acid (1:1 to 1:10), different initial amount of enzyme, etc. They have found a final conversion of over 65%. A similar work was done by Trubiano et al. [13], where the esterification of oleic acid was studied in the presence of ethanol. The authors have done a wide open screening of the operational conditions and have found the highest final conversion for a T = 45°C, a initial water concentration of 0.2% wt, a catalyst concentration of 10% and a molar ratio of 10. With these conditions the final yield was over 90% after 5 hours. The authors have also studied the possible mass transfer limitations, founding that the external mass transfer has an effect on the initial reaction rate. Tai and Brunner [15] studied the same catalyst for the esterification of palmitic acid with sugar and have found that this catalyst performed properly when using also a mixture of acetone and CO_2 in order to have supercritical pressures.

All the previous work have showed experimental data on the esterification reaction, however, it is also important the kinetics of this reaction and how it can be considered due to the fact that enzymes are a living organism and the conventional heterogeneous kinetics mechanism might not be valid or completely accurate.

Several works have been done regarding kinetics models for this reaction, some of them have been modified from established alternatives due to facts being seen on the experimental data. However, we would like to establish first the 3 kinetics model more used and well known for this type of catalyst, and then, make a short notice to some alternatives that could be found in the literature. Garcia et al. [11] have described in a very schematic way the three main kinetics mechanism:

- Random order mechanism
- Ping-Pong Mechanism
- Ordered Bi-Bi Mechanism
- King-Altman procedure[1]

Depending on how the formation of the bounds is in the reaction, some authors [16] considered that the random order mechanism could be neglected.

$$r_{m,n}=\frac{(k_{1;m,n}k_{2;n}k_{3;m}k_{4;m}[Es_{m,n}][W]-k_{-1;m,n}k_{-2;n}k_{-3;m}k_{-4;m}[Al_n][Ac_m])*[E]_{tot}\prod_{j=1}^{J}\prod_{i=1}^{I}\frac{(k_{-2;j}k_{-1;i,j}(k_{-3;i}+k_{4;i})[Al_j]+k_{3;i}k_{4;i}(k_{-1;i,j}+k_{2;j})[W])}{(k_{-2;n}k_{-1;m,n}(k_{-3;m}+k_{4;m})[Al_n]+k_{3;m}k_{4;m}(k_{-1;m,n}+k_{2;n})[W])}}{\begin{array}{l}\sum_{n=1}^{J}\sum_{m=1}^{I}\{(k_{-2;n}k_{-1;m,n}(k_{-3;m}+k_{4;m})[Al_n]+\\ k_{3;m}k_{4;m}(k_{-1;m,n}+k_{2;n})[W]+\\ k_{-2;n}k_{1;m,n}(k_{-3;m}+k_{4;m})[Es_{m,n}][Al_n]+\\ k_{1;m,n}k_{3;m}k_{4;m}[Es_{m,n}][W]+\\ k_{-4;m}k_{-3;m}k_{-2;n}[Al_n][Ac_m]+\\ k_{1;m,n}k_{2;n}(k_{-3;m}+k_{4;m})[Es_{m,n}]+\\ k_{-3;m}k_{-4;m}(k_{-1;m,n}+k_{2;n})[Ac_m]+\\ k_{-4;m}k_{3;m}(k_{-1;m,n}+k_{2;n})[W][Ac_m]+\\ k_{1;m,n}k_{2;n}k_{3;m}[Es_{m,n}][W]+\\ k_{-4;m}k_{-2;n}k_{-1;m,n}[Al_n][Ac_m])*\prod_{j=1}^{J}\prod_{i=1}^{I}\frac{(k_{-2;j}k_{-1;i,j}(k_{-3;i}+k_{4;i})[Al_j]+k_{3;i}k_{4;i}(k_{-1;i,j}+k_{2;j})[W])}{(k_{-2;n}k_{-1;m,n}(k_{-3;m}+k_{4;m})[Al_n]+k_{3;m}k_{4;m}(k_{-1;m,n}+k_{2;n})[W])}\}\end{array}} \quad (1)$$

[1] This is not a mechanism, but it will presented here.

Paiva et al. [17] have also compared some of the mechanism involved, The Ping-Pong mechanism as well as the King-Altman procedure. They have done a very good recompilation of this mechanism as well as showing different scenarios from several other works were this types of kinetics have been involved to correlate experimental data. However, it is important to notice that this expression is quite different from the conventional heterogeneous one, for that matter, it will be showed in equation 1 how a Ping Pong Bi Bi mechanism expression looks like and how much it is different from heterogeneous kinetics approaches [1,17].

When analyzing this expression several considerations need to be taking into accounts that are not necessary taking place when using other catalysts. Enzymes could get deactivated due to the reaction itself, some enzymes might required amounts of water to start being actives, however, large number of water normally deactivate the catalyst, therefore this parameters needs to be considered. Several steps might be involved regarding how the enzyme interacts and how compounds get attach to the catalyst, not all the compounds have affinity with the enzyme and therefore intermediate compounds could be produce.

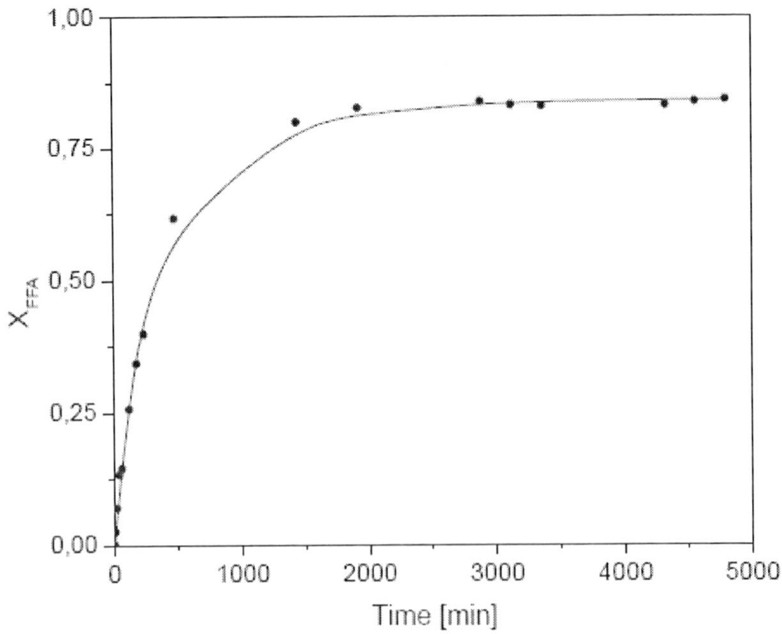

Figure 3. Fatty acid conversion using enzymes.

All this, and more other assumptions, produce that the kinetics for an enzymatic (biological catalyst) reaction are not as simple as for other alternatives catalyst. For example, Garcia et al. [11] have considered that the reaction takes place in two main reaction steps due to the fact that enzymes could be inhibited by the presence of the alcohol as well as the ester, therefore the need of some inhibition reaction terms in the main reaction rate expression. Similar results, and with the Ping-Pong Bi Bi Mechanism with competitive inhibition, were obtained by Zaidi et al. [18], whom have studied the esterification of long chain fatty acids and butanol in the presence of hexane as co-solvent.

Marchetti and Errazu [19] have done a comparison of several heterogeneous catalysts for the direct esterification of oleic acid using heterogeneous catalyst. They have found that enzymes were a great candidate to carry on the reaction, reaching conversion over 90% in middle operational conditions. Since this was a good promising results, and it is quite important to be able to have a catalyst that could be use for less pure raw material. Marchetti [20] used this enzyme for the esterification reaction of oleic acid but from an initial mixture of triglycerides and fatty acid, being as an acid oil. The results, presented in Figure 3 shows that the reaction took place and the final conversion achieved is high, over 80%, allowing this catalyst to be used as a candidate for in depth studies of reactivity.

Most important for the biodiesel industry is the transesterification reaction of oil in the presence of enzymes. Next section will be focus on this reaction.

3. Transesterification Reaction

The main raw material for biodiesel production, as established previously, is vegetable oil. Being the transesterification reaction one of the most used alternatives for producing this biofuel. We will not say that is the best or the only one, but it is, so far, the most used one. Several other options such as the hydrolysis of triglycerides and then the esterification of that fatty acid are a new option for biodiesel production. However, worldwide, transesterification has been the most studied and is the most use reaction. Being enzymes a possible catalyst to help this reaction taking place.

In this first part we would like to introduce some related work to the field, more as a review of the worked previously done and what were the main variables and results. The interaction of enzymes with the vegetable oil will be left more for the kinetics study that will be presented in the following section.

This is because the interaction might vary depending on several assumptions and conditions and therefore it will be easier to comprehend it in the following section.

Bajaj et al. [21] have done a very complete review on the transesterification reaction using enzymes. They have compared several authors and discussed not only different raw materials but also different enzymes and the operational conditions where each of them has given the best results on the biodiesel yield. They have shown that almost all of them have a yield over 90% for each case scenario, some exceptions are there but in all cases the yield was over 70%, meaning that at some point this could be improve by adding alcohol in different steps [22]. A similar work was done by Ranganatham et al. [23] whom carried on a review of several works about biodiesel produced by enzymatic catalysis. The authors have compared several works considering not only the vegetable oil and the enzyme used but also the possible cost of production. Another up to date review on this topic has been done by Antczak et al. [24]. They have compared several authors understanding and seeing what tendencies the variation on the operational variable has over the main biodiesel yield. Showing that in the mayor cases the production cost is really high. Antczak et al. [24] have also done a good review of enzymatic ways of producing biodiesel. They have summarized the lipase-catalyst transesterification for several alcohols, several fats and different enzymes. They have found that, for example, for methanol, the best yield (94.8%) was obtained after 8 hours of reaction when tallow oil and Mucor miehei IM 60 was used, but in the presence of hexane as co-solvent. Similar results have been obtained for ethanol but in 5 hours. From all the studied the authors have analyzed, the best yield was obtained for iso-butanol and tallow fat in the presence of hexane after 5 hours of reaction. The final yield was 98.5%

Dizge et al. [25] have done research on the transesterification of canola oil using Thermomyces lanuginosus immobilized on a polyglutaraldhehyde activated styrene-divinylbenzene copolymer. A yield of 97% was obtained at a 50°C reaction temperature in 24 hours reaction time. The molar ratio was 1:1 in order to not denaturalize the enzyme; however, a required ratio of 3:1 is needed in order to achieve full conversion, therefore two additions of methanol have been done during the experiments. Other works using several other enzymes could be cited, such as the work done by Tamalampudi et al. [26] where Rhizopus oryzae have been used for the reaction of Jathropha seeds and methanol. In this case the final yield was 80% after almost 3 days of reaction. They have compared their result with the same process but using a commercial enzyme and it has been found that it was required more than one extra day to reach a similar final yield. Using almost the same raw material but with another enzyme, Rathore and Madras [27] studied how good the enzyme Pangomia pinnata was for the transesterification of edible oils. They have compared this enzymatic case with supercritical alcohol reaction. They have found that to reach 70% yield was reached when the reaction took for 8 h when using enzymes, while to reach complete conversion, for supercritical alcohols, it was needed no more than 40 minutes. Two more types of enzymes were used by Noureddini et al . [28], and Royon et al. [29], being them Pseudomonas cepacia and Candida antartica. When using the process with candida antartica, the final yield was over 95% while for the other case, it was over 60%. However, it is important to point out that different operational conditions have been used and this could also lead to different results.

As when studied the esterification reaction, commercial enzymes are also available for transesterification reaction, being one of the most used the Novozym 435, being this one also good for the esterification reaction. Several works have been done using this catalyst [30-34]. Some of them are presented here to compare the different results, such as operational conditions as well as final yield. Among the works done with commercial Novozym 435, it could be found the work done by Hernandés-Martín and Otero [30] who have studied the transesterification of several vegetable oils with several alcohols in the presence of two commercial enzymes. They have found that Lipozyme TL IM produce a much faster reaction than Novozym. However the latest has been used for nine cycles, appearing to be a much more stable catalyst. Even more, Lipozyme is faster but does no reach a high final yield as it does Novozym for each tested cycle. Similar result was obtained by Maceiras et al [31], they have obtained a final yield close to 905 but in 4 h of reaction. They have used waste oil and methanol in order to study the reaction, the composition of that raw material could be seen in Table 1, extracted from reference [31]. Rodriguez et al. [32] have carried studies using soybean, sunflower, and rice bran as oil and it has been tested three enzymes, commercial ones: Novozym 435, Lipozyme TL-IM and Liposome RM-IM. It was found that Novozym 435 has the highest yield when a molar ratio of 5:1 was used. The temperature used was 30°C, and enzyme load of 15% and a reaction time of 6 h. However, the highest yield obtained was below 60%. It is important to notice that no extra additions of alcohol or catalyst have been done to the system. In a very different scenario and with different results it can be found the work done by Dalla Rosa et al. [33]. They have studied the production of biodiesel using Novozym 435 as catalyst with soybean oil and compressed propane. The results showed that complete conversion was achieved in 6 hours of reaction at 65°C and 50 bar; the reaction was carried on with ethanol. However the great yield achieved, the compressed propane system has added new cost to the general production system due to the high pressure required, contrarily the no need of extra methanol and enzyme to be added to the process make it more convenient. Hajar et al. [34], on the other hand, have used a solvent free system for their studied. They have found that a yield of 97% was possible to obtained and maintain it for several cycles completing 432 hours.

Several others works can be found in the literature where others raw material, such as waste or non edible alternatives, as well as other alcohol: ethanol, ethanol 96°, i-propanol, as well as several different enzymes and operational conditions, different reaction medium such as ionic liquids as well as combinations of acid/enzymatic alternatives, can be found [35-41]. Each of them will show advantages and restriction; proving that enzymatic catalysis is a very interested and competitive technological alternative to produce biodiesel. However, the cost of production of enzymes as well as the cost of biofuel production with this catalyst is the mayor drawback of this technology.

Table 1. Composition of the raw material [31]

Component	Composition [wt%]
C6:0	0.1
C8:0	0.5
C12:0	0.1
C14:0	0.5
C16:0	9.5
C16:1	0.9
C17:0	1.0
C18:0	4.6
C18:1	54.1
C18:2	26.6
C18:3	0.4
C20:0	0.3
C20:1	0.3
C20:2	0.4
C22:20	0.8

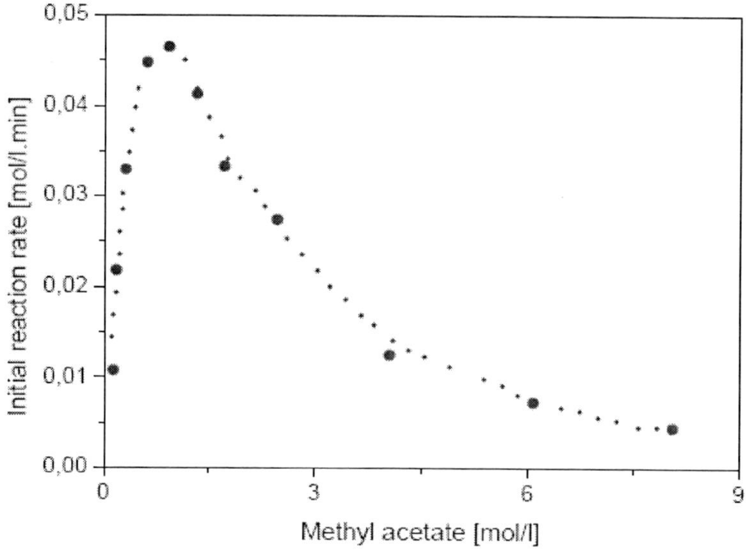

Figure 4. Comparison between model and experimental data.

4. Transesterification Kinetics

The esterification kinetics will not been discuss independently as it was done in previous chapter since it was part of the esterification reaction section. For the transesterification reaction with enzymes, different interaction mechanism has been proposed in the literature; therefore, different kinetics expression has been developed. Al-Zuhair et al. [37] have proposed a mechanism on how the interaction of triglycerides and enzymes will end up producing FAME. The mechanism could be divided into 4 parts:

- Nucleophilic addition between the enzyme and the substrate.
- Proton transfer from the acid to the alkyl oxygen atom.
- The oxygen from the methanol is added to the carbon atom that is forming a C=O bond.
- The enzyme oxygen atom of the complex is eliminated and a proton is transferred from the acid.

These four steps are well represented by Al-Zuhair et al.[37] in their work, and we advise the reader to go to this paper in order to have a much better understanding of the mechanism itself.

In their work the authors proposed a six step mechanism and two inhibition steps in order to take into account the alcohol inhibition of the enzyme and the deactivation of the catalyst due to the acylated enzyme complex.

The above six step mechanism are in good accuracy with previously published work where kinetics expression were developed, specially with the Ping-Pong Bi Bi Mechanism, as described by several authors [42-47]. According to the mechanism proposed, they have found the following kinetic expression:

$$v = \frac{V_{max}}{1+\left(K_{IA}/[A]\right)\left[1+\left([S]/K_S\right)\right]+\left(K_S/[S]\right)\left[1+\left([A]/K_A\right)\right]} \quad (2)$$

where A and S are the substrate and the alcohol concentration, K_A and K_S correspond to the dissociation constants for substrate and alcohol and V_{max} is the maxim reaction rate possible.

If ethanol is used and it is added to the system in a continuous feed in order to maintain the molar ratio, the final equation is more complicated, as described by Pessoa et al. [47]. It is important to take into account that the kinetic parameters were fit in relation with the experimental data, and for those data the result is quite satisfactory. However, if other data is used, despite the fact that it might follow a Ping-Pong Bi Bi mechanism, the parameters should be re adjusted. If instead of an alcohol, a methyl acetate is used to carry on the transesterification reaction [45] (in this case is called interesterification reaction) the final kinetic expression is very similar to 2. However, due to the fact that other compounds are involved, and therefore other mechanism step might be the ones to follow, there is a slight modification in the final expression, being as follows:

$$v_i = \frac{V_{max}[TG][A]}{K_{mTG}[A]\left[1+\left([A]/K_i\right)\right]+[TG]K_{mA}+[TG][A]} \quad (3)$$

This expression is also based on a Ping Pong Bi Bi mechanism with competitive substrate inhibition. With this expression, and considering that the concentration of triglycerides and methyl acetate are independent, the authors have compared experimental data of initial reaction rate for several concentration of methyl acetate with the model proposed and have found a very good agreement as shown in Figure 4. Extracted from reference [47].

However, in order to obtain a simpler mathematical expression for the variation of triglycerides, monoglycerides, etc. they have proposed a very simple three elementary steps taking place, as described for the previous mechanism as well. Using those simple steps the authors were able to obtained mathematical expression for all the compounds, as follow:

$$\begin{aligned}
&\frac{d[TG]}{dt} = -k_1[TG][A] + k_2[DG][ME] \\
&\frac{d[TG+DG]}{dt} = -k_3[DG][A] + k_4[MG][ME] \\
&\frac{d[TG+DG+MG]}{dt} = -k_5[MG][A] + k_6[TA][ME] \\
&[TG]_0 = [TG] + [DG] + [MG] + [GL] \\
&3[TG]_0 = 3[TG] + 2[DG] + [MG] + [ME] \\
&[A]_0 = [ME] + [A]
\end{aligned} \tag{4}$$

Even though this simpler kinetics model satisfactory reproduce the experimental data, as it was presented, more complicated interaction between enzymes and triglycerides are likely to be taking place. However, this is not a very easy task and it has not been extremely addressed since this is a quite complicated problem. Al-Zuhair [48] has done some kinetic study for the interaction of triglycerides with alcohol to produce biodiesel. He has extrapolated the theory behind the Ping Pong mechanism for the esterification and for the transesterification. The author has found a new model that shows good agreement with the experimental data. This tendency of more complex mechanism could be improved towards a more realistic but much more complex kinetics mechanism, allowing a better detailed description of the process.

5. Economics

As previously showed, this is a very important part of any business decision; therefore, a very detail study should be done in order to see how possible and how profitable a technology is. For enzymes, it has been pointed out that a good enzyme that has a good activity life time has a so high price that will make process non profitable when analyzing the whole plant with a very high degree of detail in each part involved. Nevertheless, it should not be rule out completely, since every day new enzymes, new process, new substrate are being tested, and it is possible that a new option will make enzymes a more competitive technology.

References

[1] Marchetti, J.M., Miguel, V.U., Errazu, A.F. *Possible methods for biodiesel production. Renewable and Sustainable Energy Reviews*. 11. (2007), 1300-1311.

[2] Perez G.. Master Thesis *Analysis of enzymatic alcoholisms reaction with vegetables oils*. February 2003.

[3] Drapcho, C.M., Nhuan, N.P., Walker, T.H., *Biofuels Engineering Process Technology*. Mc Graw Hill Companies, Inc. 2008.

[4] Balat, M., Balat, H., Öz, C. *Progress in bioethanol processing. Progress in Energy and Combustion Science*. 34. (2008), 551-573.

[5] McMillan, J.D. *Bioethanol Production: Status and Prospects. Renewable Energy*. 10(2-3). (1997), 295-302.

[6] Keshwani, D.R., Cheng, J.J. *Switchgrass for bioethanol and other value added applications: A review. Bioresource Technology*. 100. (2009), 1515-1523.

[7] Binod, P., Sindhu, R., Singhania, R.R., Vikram, S., Devi, L., Nagalakshmi, S., Kurien, N., Sukumaran, R.K., Pandey, A. *Bioethanol production from rice straw: an overview. Bioresource Technology*. (2009), (article in press)

[8] Alvira, P., Tomás-Pejó, E., Ballesteros, M., Negro, M.J. *Pretreatment technologies for an efficient bioethanol production process based on enzymatic hydrolysis: A review. Bioresource Technology*. (2009), (article in press)

[9] Eggeman, T., Elander, R.T. *Process and economic analysis of pretreatment technologies. Bioresource Technology*. 96. (2005), 2019-2025.

[10] Dashtan, M., Schraft, H., Qin, W. *Fungal bioconversion of lignocellulosic residues; opportunities and perspectives. International Journal of Biological Science*. 5(6). (2009), 578-595.

[11] Garcia, T., Sanchez, N., Martinez, M., Aracil, J. *Enzymatic synthesis of fatty esters part I: kinetic approach. Enzyme and Microbial Technology*. 25. (1999), 584-590.

[12] Torres, C., Otero, C. *Part III. Direct enzymatic esterification of lactic acid with fatty acids. Enzyme and Microbial Technology*. 29. (2001), 3-12.

[13] Trubiano, G., Borio, D., Errazu, A. *Influence of the operating conditions and the external mass transfer limitations on the synthesis of fatty acid esters using a Candida antartica lipase. Enzyme and Microbial Technology*. 40. (2007), 716-722.

[14] Quintana, P.G., Baldessari, A. *Lipase-catalyzed regioselective preparation of fatty acid esters of hydrocortisone. Steroids*. 74. (2009), 1007-1014.

[15] Tai, H.P., Brunner, G. *Sugar fatty acid ester synthesis in high pressure acetone-CO_2 systems. Journal of Supercritical Fluids*. 48. (2009), 36-40.

[16] Brzozowski, A.M., Derewenda, U., Derewenda, Z.S., Godson, G.G., Lawson, D.M., Turkenburg, J.P., Bjorkling, F., Huge-Jensen, B., Patkar, S.A., Thim, L. *A model for interfacial activation in lipases from the structure of a fungal lipase-inhibitor complex. Nature*. 351. (1991), 491-494.

[17] Paiva, A.L., Balcão, V.M., Malcata, F.X. *Kinetics and mechanisms of reactions catalyzed by immobilized lipases. Enzyme and Microbial Technology*. 27. (2000), 187-204.

[18] Zaidi, A., Gainer, J.L., Carta, G., Mrani, A., Kadiri, T., Belarbi, Y., Mir, A. *Esterification of fatty acids using nylon-immobilized lipase in n-hexane: kinetic parameters and chain-length effects. Journal of Biotechnology*. 93. (2002), 209-216.

[19] Marchetti, J.M., Errazu, A.F. *Comparison of different heterogeneous catalysts and different alcohols for the esterification reaction of oleic acid. Fuel*. 87. (2008), 3477-3480.

[20] Marchetti, J.M. Ph.D. thesis. *Technological alternatives for biodiesel production*. Universidad Nacional del Sur. Argentina (2008).

[21] Bajaj, A., Lohan, P., Jha, P.N., Mehrotra, R. *Biodiesel production through lipase catalyzed transesterification: a review. Journal of Molecular Catalysis B: Enzymatic*. 62. (2010), 9-14.

[22] Shimada, Y., Watanabe, Y., Sugihara, A., Tominaga, Y. *Enzymatic alcoholysis for biodiesel fuel production and application of the reaction to oil processing. Journal of Molecular Catalysis B: Enzymatic*. 17. (2002), 133-142.

[23] Ranganathan, S.V., Narasimhan, S.L., Muthukumar, K. *An overview of enzymatic production of biodiesel. Bioresource Technology*. 99. (2008), 3975-3981.

[24] Antczak, M.S., Kubiak, A., Antczak, T., Bielecki, S. *Enzymatic biodiesel synthesis - key factors affecting efficiency of the process. Renewable Energy*. 34. (2009), 1185-1194.

[25] Dizge, N., Aydiner, C., Imer, D.Y., Bayramoglu, M., Tanriseven, A., Keskinler, B. *Biodiesel production from sunflower, soybean and waste cooking oils by transesterification using lipase immobilized onto a novel microporous polymer. Bioresource Technology*. 100. (2009), 1983-1991.

[26] Tamalampudi, S., Talukder, M.R., Hama, S., Numata, T., Kondo, A., Fukuda, H. *Enzymatic production of biodiesel from Jatropha oil: A comparative study of immobilized-whole cell and commercial lipases as a biocatalyst. Biochemical Engineering Journal*. 39. (2008), 185-189.

[27] Rathore, V., Madras, G. *Synthesis of biodiesel from edible and non-edible oils in supercritical alcohols and enzymatic synthesis in supercritical carbon dioxide. Fuel*. 86. (2007), 2650-2659.

[28] Noureddini, H., Gao, X., Philkana, R.S. *Immobilized Pseudomonas cepacia lipase for biodiesel fuel production from soybean oil. Bioresource Technology*. 96. (2005), 769-777.

[29] Royon, D., Daz, M., Ellenrieder, G., Locatelli, S. *Enzymatic production of biodiesel from cotton seed oil using t-butanol as a solvent. Bioresource Technology*. 98. (2007), 648-653.

[30] Hernandéz-Martín, E., Otero, C. *Different enzyme requirements for the sy thesis of biodiesel: Novozym 435 and Lipozyme TL IM. Bioresource Technology*. 99. (2008), 277-286.

[31] Maceiras, R., Vega, M., Costa, C., Ramos, P., Márquez, M.C. *Effet of methanol content on enzymatic production of biodiesel from waste frying oil. Fuel*. 88. (2009), 2130-2134.

[32] Rodrigues, R.C., Volpato, G., Wada, K., Ayub, M.A.Z. *Enzymatic synthesis of biodiesel from transesterification reactions of vegetble oils*

and short chaing alcohols. Journal of American oil Chemists Society. 85. (2008) 925-930.

[33] Dalla Rosa, C., Morandim, M.B., Ninow, J.L., Oliveira, D., Trichel, H., Vladimir Oliveira, J. *Lipase-catalyzed production of fatty acid ethyl esters from soybean oil in compressed propane. The Journal of Supercritical Fluids*. 47. (2008), 49-53.

[34] Hajar, M., Shokrollahzadeh, S., Vahabzadeh, F., Monazzami. A. *Solvent free methanolysis of canola oil in a packed bed reactor with use of Novozym 435 plus loofa. Enzyme and Microbial Technology*. 45. (2009), 188-194.

[35] Lozano, P., De Diego, T., Carrié, D., Vaultier, M., Iborra, J.L. *Enzymatic ester synthesis in ionic liquids. Journal of Molecular Catalysis B: Enzymatic*. 21. (2003), 9-13.

[36] Ting, W.J., Huang, C.M., Giridhar, N., Wu, W.T. *An Enzymatic/acid-catalyzed hybrid process for biodiesel production from soybean oil. Journal of Chinese Institute of Chemical Engineering*. 39. (2008), 203-210.

[37] Al-Zuhair, S., Ling, F.W., Jun, L.S. *Proposed kinetic mechanism of the production of biodiesel from palm oil using lipase. Process Biochemistry*. 42. (2007), 951-960.

[38] Nielsen, P.M., Brask, J., Fjerbaek, L. *Enzymatic biodiesel production: technical and economical considerations. European Journal of Lipids Science and Technology*. 110. (2008), 692-700.

[39] Akoh, C.C., Chang, S.W., Lee, G.C., Shaw, J.F. *Enzymatic approach to biodiesel production. Journal of Agricultural and Food Chemistry*. 55(22). (2007), 8995-9005.

[40] Ha, S.H., Lan, M.N., Lee, S.H., Hwang, S.M., Koo, Y.M. *Lipase-catalyzed biodiesel production from soybean oil in ionic liquids. Enzyme and Microbial Technology*. 41. (2007), 480-483.

[41] Chen, Y., Xiao, B., Chang, J., Fu, Y., Ly, P., Wang, X. *Synthesis of biodiesel from waste cooking oil using immobilized lipase in fixed bed reactor. Energy Conversion and Management*. 50. (2009), 668-673.

[42] Janssen, A.E.M., Vaidya, A.M., Halling, P.J. *Substrate specificity and kinetics of Candida rugosa lipase in organic media. Enzyme and Microbial Technology*. 18. (1992), 340-346.

[43] Rizzi, M., Stylos, P, Riek, A., Reuss, M. *A kinetic study of immobilized lipase catalyzing the synthesis of isoamyl acetate by transesterification in n-hexane. Enzyme and Microbial Technology*. 14. (1992), 709-714.

[44] Marty, A., Chulalaksananukul, W., Willemot, R.M., Condoret. J.S. *Kinetics of lipase-catalyzed esterification in supercritical CO_2.* *Biotechnology and Bioenergineering*. 39. (1992), 273-280.

[45] Xu, Y., Du, W., Liu, D. *Study on the kinetics of enzymatic interestification of triglycerides for biodiesel production with methyl acetate as the acyl acceptor. Journal of Molecular Catalysis B: Enzymatic.* 32. (2005), 241-245.

[46] Torres, C.F., Vázquez, L., Señoráns, F.J., Reglero, G. *Enzymatic síntesis of shor chain diacylated alkylglycerols: A kinetic study. Process Biochemistry.* 44. (2009), 1025-1031.

[47] Pessoa, F.L.P., Magalhães, S.P., Falcão, P.W.C. *Kinetic study of biodiesel production by enzymatic transesterification of vegetable oil.* 10th International Symposium on Process System Engineering-PSE2009. (2009), 1809-1814.

[48] Al-Zuhair, S. *Production of Biodiesel by Lipase catalyzed transesterification of vegetable oils: a kinetics study. Biotechnology Progress.* 21. (2005), 1442-1448.

Chapter 8

SUPERCRITICAL TECHNOLOGY

1. INTRODUCTION

The use of supercritical compounds it is well established and has been used for several chemical applications such as essential oil extraction from vegetable plants [1-5] as well as many other applications [6]:

- Dry cleaning
- Supercritical fluid chromatography
- Chemical Reactions
- Impregnation and dyeing
- Nano and micro particle formation
- Supercritical drying.
- Supercritical water oxidation
- Biodiesel production
- Supramics
- Refrigeration

The supercritical process consists on reaching a certain operational conditions that will allow one of the fluids to be in its supercritical state. If we considered any compound and we look at a P and T plot (pressure and temperature) we can see different regions, solid, liquid, vapor and a supercritical area, see Figure 1. Extracted from reference [7].

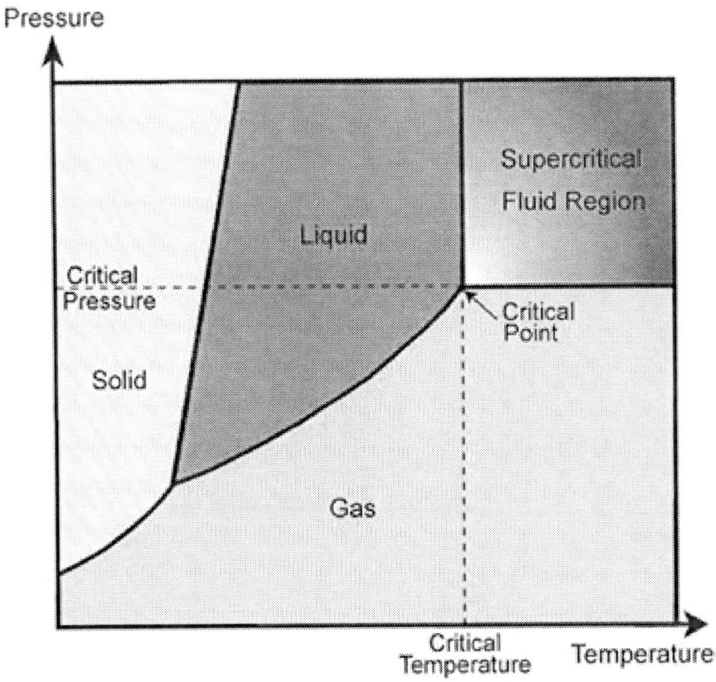

Figure 1. PT diagram.

When reaching a certain pressure and temperature, both, liquid and gas phase becomes one and there is not possible distinction among them, therefore we have reached the supercritical pressure and temperature and if any of this continues to increases we will be in the supercritical region.

This option has become more and more of interest on the biodiesel community, as it will be point out, this option has very short reaction times and it is suitable to be used with almost any raw material, allowing it to be very promising for waste or frying oils. Even more, with this technology no catalyst is required. [8-11]

However, more drastic operational conditions, high pressure (over 200 bar) and high temperatures (could be over 200°C), are required to obtained a supercritical phase with the alcohol and also to have the conditions that will allow the transesterification reaction to take place and been carried on [12-13].

Contrarily to other alternatives, this option presents some advantages on the down streaming effluents and over the reaction time. Since there is no catalyst involved, the separation after the reactor is much simpler and besides reducing the pressure, a simple decanter might be a good choice to produce the separation of the glycerol from the biodiesel phase. Nevertheless, a distillation column might be suitable before this equipment in order to extract the excess of methanol used to carry on the reaction.

As mentioned, since this technology allows the treatment of almost any raw material, the study of both reactions is of high interested. However, it is common to find both reactions being studied simultaneously and not independently as done for other technologies.

2. ESTERIFICATION REACTION

Due to the fact, as it will be presented in the next section, that supercritical alcohol could treat both fatty acid and triglycerides indistinctly; most of the work in the literature is related to either transesterification or both reactions simultaneously. Most of the time it is shown that the presence of fatty acid does not have a negative effect on the biodiesel production as it could be the case when using conventional basic homogenous approach.

However, we would like to make a few comments in this reaction on its own before addressing the transesterification reaction. This is because fatty acid could be part of the raw material, as in a crude oil described in Chapter 3, but also because it is possible to obtain them from the hydrolysis of triglycerides, according to the following reaction [14]

$$
\begin{array}{lclclcl}
CH_2\text{-}COO\text{-}R_1 & & & & CH_2\text{-}OCOCH_3 & & R_1\text{-}COOH \\
| & & & & | & & \\
CH\text{-}COO\text{-}R_2 & + & 3CH_3\text{-}COOH & \leftrightarrow & CH\text{-}OCOCH_3 & + & R_2\text{-}COOH \\
| & & & & | & & \\
CH_2\text{-}COO\text{-}R_3 & & & & CH_2\text{-}OCOCH_3 & & R_3\text{-}COOH \\
\text{Triglycerides} & & \text{Acetic acid} & & \text{Triacetin} & & \text{Fatty acid}
\end{array}
\qquad (1)
$$

When the fatty acids are produced, they can be esterified by any technique into biodiesel. Supercritical approach is also suitable to do this as described by Saka et al. [14] where rapeseed oil were hydrolyzed into fatty acid in a subcritical reaction for a further supercritical esterification of the fatty acids with methanol. The authors tested several process variables for the hydrolysis reaction in order to improve its conversion to fatty acid. For the best scenario, the produced fatty acid were esterified into biodiesel under high pressure (20 MPa) and high temperature (300°C) in order to reach high conversion in a few minutes. The authors proposed kinetics for the conversion of TG into FFA, they have proposed a simple first order kinetics related only to the triglycerides. They have found good agreement between the experimental data and the proposed model.

A similar work, done by Minami and Saka [15] shows that the hydrolysis and further esterification could be done in the presence of water instead of acetic acid. For this work they have used rapeseed oil and have tested several conditions as well. However, in this case the kinetics mechanism consider an autocatalytic reaction for the hydrolysis, making a more complex set of mathematical equations since in this case monoglycerides and diglycerides are also produced and could also be consumed by hydrolysis. The authors have also proposed a simple second order reversible reaction for the biodiesel production, being as follow:

$$\frac{dC_{FAME}}{dt} = \left(kC_{FFA}C_M - k'C_W C_{FAME}\right)C_{FFA} \qquad (2)$$

The authors have found an excellent agreement between data and model.

Alenezi et al. [16] have proposed a similar kinetics model based on the experimental data of the esterification reaction of fatty acids with methanol in supercritical conditions. In this case, the pressure has been half of the one used by Saka et al. [14], but other variables such as molar ratio and agitation speed have been tested, giving high final conversion in very short reaction time. The kinetic model proposed is a second order reversible reaction for the esterification reaction.

As presented above, the esterification reaction is important for this technology; however, supercritical alcohol has been more used for biodiesel production by transesterification reaction of vegetable oils. Therefore, the following section will be dedicated to the transesterification of oils with supercritical alcohols.

3. Transesterification Reaction

The transesterification reaction of vegetable oils in the presence of alcohols in a supercritical state does not require a catalyst to allow the reaction to take place. However, in some cases, as it will be shown, it is possible to add either co-solvents as well as catalyst in order to reduce some of the operational conditions such as molar ratios.

When carry on this reaction, the most common alcohol is methanol, several works could be found in the literature where methanol is used in its supercritical state. Among then we can found the work done by Hawash et al. [8] have carried on the transesterification reaction using Jatropha oil and supercritical methanol. The reaction time was less than four minutes in order to achieve a yield of 100%. The reaction temperature was 320°C with an operation pressure of around 80 bar. The molar ratio was of 43:1 alcohol/oil. The amount of free fatty acid was around 2% w/w, which under the operational conditions it were esterifies into biodiesel as well. Similar work was done by Saka and Kusdiana [13] have studied the reaction of rapeseed oil with methanol. They carried the experiments with a reaction time of 4 minutes, a molar ratio of 42:1 and a pressure over 80 bar and a temperature higher than 240 C. The final yield was higher than 98% and the free fatty acids were also esterified. Tan et al. [17] studied the same reaction but using palm oil as raw material. They have found that 70% conversion was achieved after 20 minutes of reaction at a temperature of 360°C and a molar ratio of 30:1. For the transesterification of RBD palm oil with methanol, Song et al. [18] studied several operational conditions. They have found that among the tested molar ratio (3:1 to 80:1) and the temperature (200-400°C) one of the best scenario was obtained after 5 minutes of reaction at 350°C with a molar ratio of 30:1. If the molar ratio is increased, the reaction time gets reduced in less than a minute; therefore, an economic decision should be made regarding both variables. Different from the previous work, He et al. [19] have found that increasing gradually the heating of the system will produced a higher final conversion under similar conditions. The yield when using constant temperature was not higher than 77% while when using gradual heating the final yield reached was over 96%.

Besides methanol, several works have also tested ethanol as an alternative alcohol due to the fact that it might be produce from renewable sources and produce a more natural fuel. Tan et al. [20] have compared both alcohols in the same operational conditions in order to see the benefits or disadvantages of using one or the other. They have found that with ethanol the conversion achieved was almost 3% lower. A similar final conversion for ethanol as alcohol was found by Gui et al. [21]. They have performed the supercritical transesterification reaction of palm oil using ethanol. They have found that a yield of 79.2% was obtained when 349°C, 30 minutes of reaction time, a molar ration of 33:1 and a pressure over 63.8 bar were used. Another worked where both types of alcohols were used was done by Madras et al. [22]. In this work, ethanol shows a higher final conversion compared to methanol, slightly higher but higher, reaching almost 100% for ethanol and over 96% for methanol. The operational conditions are different from previous works having higher temperatures as well as pressure. Even more, the authors have compared this technology with the enzymatic process but carried out under supercritical CO_2 solvent. It was found that 30% conversion was achieved when an enzyme is used in a supercritical carbon dioxide medium. Using the same principle, Varma et al. [23] have tested enzymes under supercritical CO_2 for the transesterification reaction. In this case the final conversion was higher than in the previous work (70%) but not high enough as to be considered a possible industrial technological solution. Even more, the reaction time for this matter was considerable high, with over 20 h of reaction. However, this could lead to the fact that catalyst as well as co-solvent could be used under supercritical conditions in order to improve the reaction under study. We will try to review some works dealing with both of this new possibilities.

Some works, such as those done by Kasim et al. [24] and Cao et al. [25], carried on the supercritical transesterification reaction in the presence of a co-solvent with the aim of reducing the pressure, temperature as well as the molar ratio. Kasim et al. [24] used CO_2 as co-solvent and obtained a yield of 94.8% and 51.3% when using refine oil and rice bran respectively. The reaction was done at 300°C, 300 bar, with a molar ratio of 27:1 and a reaction time of 5 minutes. Differently, Cao et al. [25] have used propane as co-solvent, producing a more drastic reduction on the operational conditions. The final yield of over 99% was reached for a propane/methanol ratio of 0.05, a methanol/oil ratio of 24:1, pressure of 128 bar, a T = 300°C and 5 minutes of reactions.

On the other hand, it is possible to add some catalyst to the system in order to improve the reaction, allowing lower operational conditions or to increase the reaction rate. Demirbas [26] used sunflower oil in a supercritical environment of methanol; however, contrarily to others works, calcium oxide was added as catalyst. It was found that high amounts of catalyst was necessary to obtained high yield of biodiesel, while using 1% reaches very low results. Full conversion was achieved at 525 K, 41:1 of molar ratio, a concentration of catalyst of 3% and a pressure of 240 bar. However, the presence of a catalyst will produce a different and possible more complicated down streaming separation and purification. A similar work was done by the same author [27] but in this case it was used MgO instead of calcium oxide. It was found that a yield of over 90% was achieved when using a catalyst.

However, the reaction temperature has a very important effect on the conversion achieved, needing a temperature over 200°C to reach a good final yield.

So far we tried to show the different approaches available to carry on the transesterification reaction while using the alcohol in its supercritical state. It is also important to be able to obtained a mathematical model of this reaction, since the process has changed from the previous alternative it is very likely that the kinetics involved are also different, due to the fact that there is not catalyst but also to the high pressure and temperatures involved. This last fact it is also a key factor in the economic evaluation of this technology, the investment on equipment to be able to handle this conditions as well as the operational cost to achieved them could be so significant that the process could be economically unattractive. Because of this, we will like to address the kinetics as well as the economic study for this technology.

4. Kinetics Model for the Transesterification Reaction

As previous discussed for all the technologies from Chapter 4 to Chapter 7, the mechanism involved and the reactions considered as elementary are the key factor when establish a reaction mechanism path for the triglyceride to be modified into esters.

Based on the overall reaction:

$$TG + 3CH_3OH \quad \leftrightarrow \quad 3RCOOCH_3 + Glycerol \qquad (3)$$

A very simple kinetics model was developed by different authors: Kusdiana and Saka [9], and He et al. [28]. In all those works the expression 3 was considered to not be reversible based that the amount of alcohol is quite big (42:1 molar ratio), then the reaction will be displace towards the products. Even more, it is assumed that the alcohol concentration does not change during the reactions due to its large amount. Base on those assumptions, the variation of the triglycerides over time is due only to the concentration of unreacted material (to include diglycerides as well as possible monoglycerides), so the reaction rate expression will be as follows, defined by Kusdiana and Saka [9], and He et al. [28]:

$$\frac{d[uME]}{dt} = -k[uME] \quad (4)$$

This expression was satisfactory employed by Kusdiana and Saka [9],and He et al. [28] to correlate their experimental data. The kinetics parameters obtained for the reaction constant for both experimental data is quite similar considering that different raw materials and slightly modified operational conditions were used.

In order to take into account more variables in the kinetic model and to proposed a little more complicated mechanism that could represent to other extend what is happening in the reactor, Song et al. [18] proposed that the main reaction taking place is the overall reaction (expression 3), but the variations on the TG concentration it is due to changes in the triglycerides concentration as well as in the alcohol concentration. This introduces a new variable that has been not assumed constant in the previous models.

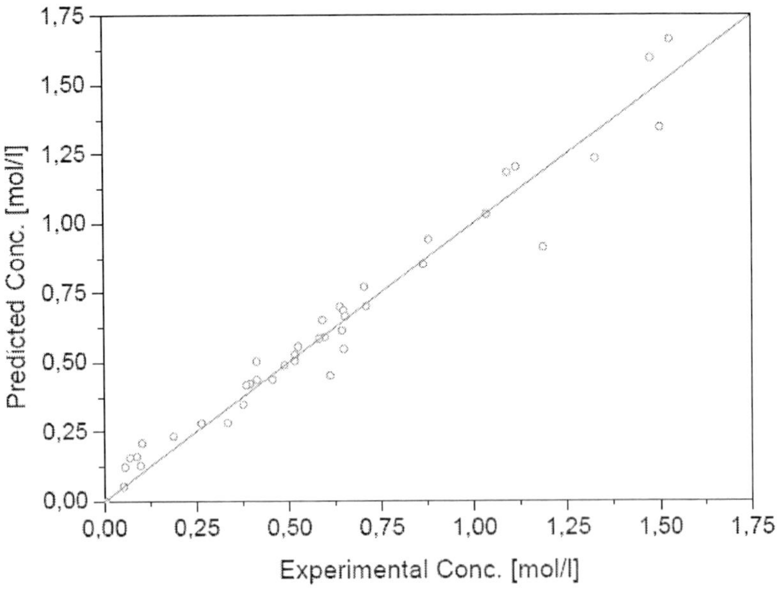

Figure 2. Comparison of experimental data with predicted. Extracted from [18].

With this new hypothesis the authors obtained expression 5, where it can be seen that both concentrations are relevant for the study of the time depend variation of triglycerides concentration.

$$\frac{dC_{TG}}{dt} = -\left(4.3376 \bullet 10^5\right)\exp\left(-\frac{1.0527 \bullet 10^5}{RT}\right)C_{TG}^{0.9565}C_{MeOH}^{1.0493} \tag{5}$$

In order to justify this new model, Figure 2 shows the comparison between the estimated concentrations of biodiesel with the experimental value. Figure 2 shows an excellent agreement between both data verifying this alternative kinetic model.

5. Economic Study for Supercritical Technology

As point out by all the previous worked, it is suitable to carry on the reaction under supercritical conditions, normally with methanol as reactant. Although it is physical possible, the commercial part of this technology is still under study due to the high investment cost and the high operational cost associated to high pressure and temperature.

Several economic studies have been done in order to establish the feasibility of this alternative. Van Kasteren and Nisworo [29] have studied different scenarios for biodiesel production through a supercritical process, being the main difference the plant scale and the production flow of biofuel. They have found that this process could be competitive with conventional technology, despite the high investment cost. Similar results have been obtained by Deshpane et al. [30] where they have studied the production of biofuel but with a power cogeneration, allowing this to have a new perspective of the market and of the most relevant variables of the process. The inclusion of the cogeneration has allow a small production plant (2424 gal/day of biodiesel) to be economically attractive, while Van Kasteren and Nisworo [29] have reached that the most profitable scenario was related to 125.000 tonnes/year.

Marchetti and Errazu [31] have done a similar work studying the economic feasibility of a supercritical plant; however, they have studied a process involving two reaction step with separation in between. With these new assumptions and the prices according to the market of where the plant would have been installed, the economic analysis has, unfortunately, showed that the process is not economically attractive. Some improvements need to be done, such as scale up or a reduction in the amount of equipment, etc, in order to have a competitive technology. This worked was based on kinetics from other authors and experimental data from other research works, Lim et al. [32] have done a economic study comparing alkali alternatives with supercritical process from the literature as well as one with the experimental data produced in their own research facilities. They have found, for a one supercritical reactor, this technology has a lower payout time compared with an alkali alternative when the production rate is of 8000 ton/year. The total fixed capital cost for this work is in good agreement with the one estimated by Van Kasteren and Nisworo [29] for the same plant capacity.

This technology is gaining more relevance each day due to the benefits regarding reaction time and the less downstreaming separation equipment required. Even more, economic studies have shown that this technology is becoming competitive compared with conventional alternatives.

References

[1] Döker, O., Salgin, U., Yildiz, N., Aydoğmuş, M., Çalimli, A. *Extraction of sesame seed oil using supercritical CO_2 and mathematical modeling. Journal of Food Engineering.* 97. (2010), 360-366.

[2] Corso, M.P., Fagundes-Klen, M.R., Silva, E.A., Filho, L.C., Santos, J.N., Freitas, L.S., Dariva, C. *Extraction of sesame seed (Sesamun indicum L.) oil using compressed propane and supercritical carbon dioxide. Journal of Supercritical Fluids*. 52. (2010), 56-61.

[3] Fiori, L. *Supercritical extraction of sunflower seed oil: Experimental data and model validation. Journal of Supercritical Fluids.* 50. (2009), 218-224.

[4] Jia, D., Li, S., Xiao, L. *Supercritical CO_2 extraction of Plumula nelumbinis oil: Experiments and modeling. Journal of Supercritical Fluids.* 50. (2009), 229-234.

[5] Han, X., Cheng, L., Zhang, R., Bi, J. *Extraction of safflower seed oil by supercritical CO_2. Journal of Food Engineering*. 92. (2009), 370–376.

[6] www.wikipedia.com

[7] http://www.separex.fr

[8] Hawash, S., Kamal, N., Zaher, F., Kenawi, O., El Diwani, G. *Biodiesel fuel from Jatropha oil via non-catalytic supercritical methanol transesterification. Fuel*. 88. (2009), 579-582.

[9] Kusdiana, D., Saka, S. *Kinetics of transesterification in rapeseed oil to biodiesel fuel as treated in supercritical methanol. Fuel*. 80. (2001), 693-698.

[10] Demirbaş, A. *Biodiesel fuels from vegetable oils via transesterification in supercritical methanol. Energy Conversion and Management*. 43. (2002), 2349-2356.

[11] Demirbaş, A. *Production of biodiesel fuels from linseed oil using methanol and ethanol in non-catalytic SCF conditions. Biomass and Bioenergy*. 33. (2009), 113-118.

[12] Demirbaş, A. *Biodiesel fuels from vegetable oils via catalytic and non catalytic supercritical alcohol transesterifications and other methods: a survey. Energy Conversion and Management*. 44. (2003), 2093-2109.

[13] Saka, S., Kusdiana, D. *Biodiesel fuel from rapeseed oil as prepared in supercritical methanol. Fuel*. 80. (2001), 225-231.

[14] Saka, S., Isayama, Y., Ilham, Z., Jiayu, X. *New process for catalyst-free biodiesel production using subcritical acetic acid and supercritical methanol. Fuel,* (2009), article in press.

[15] Minami, E., Saka, S. *Kinetics of hydrolysis and methyl esterification for biodiesel production in two-step supercritical methanol process*. Fuel. 85. (2006), 2479-2483.

[16] Alenezi, R., Leeke, G.A., Winterbottom, J.M., Santos, R.C.D., Khan, A.R. *Esterification kinetics of free fatty acids with supercritical methanol for biodiesel production. Energy Conversion and Management*. (2010). Article in Press.

[17] Tan, K.T., Lee, K.T., Mohamed, A.R. *Production of FAME by palm oil transesterification via supercritical methanol technology. Biomass and Bioenergy*. 33. (2009), 1096-1099.

[18] Song, E.S., Lim, J.W., Lee, H.S., Lee, Y.W. *Transesterification of RBD palm oil using supercritical methanol. Journal of Supercritical Fluids*. 44. (2008), 356-363.

[19] He, H., Wang, T., Zhu, S. *Continuous production of biodiesel fuel from vegetable oil using supercritical methanol process. Fuel*. 86. (2007), 442-447.

[20] Tan, K.T., Gui, M.M., Lee, K.T., Mohamed, A.R. *An optimized study of methanol and ethanol in supercritical alcohol technology for biodiesel production. Journal of Supercritical Fluids*. 49(2). (2009), 286-292.

[21] Gui, M.M., Lee, K.T., Bhatia, S. *Supercritical ethanol technology for the production of biodiesel: process optimization studies. The Journal of Supercritical Fluids*. 49. (2009), 286-292.

[22] Madras, G., Kolluru, C., Kumar, R. *Synthesis of biodiesel in supercritical fluids. Fuel.* 83. (2004), 2029-2033.

[23] Varma, M.N., Deshpande, P.A., Madras, G. *Synthesis of biodiesel in supercritical alcohols and supercritical carbon dioxide. Fuel.* (2009) Article in press.

[24] Kasim, N.S., Tsai, T.H., Gunawan, S., Ju, Y.H. *Biodiesel production from rice bran oil and supercritical methanol. Bioresource Technology.* 100. (2009), 2399-2403.

[25] Cao, W., Han, H., Zhang, J. *Preparation of biodiesel from soybean oil using supercritical methanol and co-solvent. Fuel.* 84. (2005), 347-351.

[26] Demirbaş, A. *Biodiesel from sunflower oil in supercritical methanol with calcium oxide. Energy Conversion and Management.* 48. (2007), 937-941.

[27] Demirbaş, A. *Biodiesel from vegetable oils with MgO catalytic transesterification in supercritical methanol.* Energy Source, Part A: Recovery, Utilization and Environmental Effects. 30. (2008), 1645-1651.

[28] He, H., Sun, S., Wang, T., Zhu, S. *Transesterification kinetics of soybean oil for production of biodiesel in supercritical methanol. Journal of the American Oil Chemists Society.* 84. (2007), 399-404.

[29] van Kasteren, J.M.N, Nisworo, A.P. *A process model to estimate the cost of industrial scale biodiesel production from waste cooking oil by supercritical transesterification. Resources, Conservation and Recycling*. 50. (2007), 442-458.

[30] Deshpane, A., Anitescu, G., Rice, P.A., Tavlarides, L.L. *Supercritical biodiesel production and power cogeneration: technical and economic feasibilities. Bioresource Technology*. 101. (2010), 1834-1843.

[31] Marchetti, J.M., Errazu, A.F. *Technoeconomic study of supercritical biodiesel production plant. Energy Conversion and Management.* 49. (2008), 2160-2164.

[32] Lim, Y., Lee, H.S., Lee, Y.W., Han, C. *Design and Economic Analysis of the Process for Biodiesel Fuel Production from Transesterificated Rapeseed Oil Using Supercritical Methanol. Industrial and Engineering Chemistry Research.* 48. (2009), 5370-5378.

Chapter 9

OTHERS POSSIBLE TECHNOLOGIES

1. INTRODUCTION

In all the previous chapters we tried to focus on the technologies that have been developed over the last decades, showing you the different advantages and disadvantages as done in many review papers [1-7] but trying to add more information to each of them, showing some research work as well as kinetics models and economic studies regarding each of this alternatives.

However, there are several other options involved for biodiesel production that are gaining more and more relevance during the last years and are being developed since then. In this topic, I will like to show membrane reactors, reactive distillation as well as monolithic reactors [8-10]. I think these are very new technologies that more research could be done and new economic alternatives for biodiesel production could be produced.

Membrane reactors have been widely used for steam reforming with the aim of separation hydrogen from the main stream [11-13], water gas shift reactions; however, there are several other options involved for biodiesel production that are gaining more and more relevance during the last years and are being developed since then.

A membrane reactor consist, as see in Figure 1, of a main flow where reactants are involved and a sweep gas that will allow the product to be separate easily from the rest. The reaction could take place in the homogeneous phase, if sodium hydroxide is used, could also be considered to have the membrane reactor in a heterogeneous fix bed reactor. In this last case, the reaction is much simpler and the product much pure. However, it was proved by Dubé et al. [8] that the biodiesel produced in their experimental setup has very good quality, due to a great separation with the membrane.

In their work, Dubé et al. [8] have used a conventional catalyst (potassium hydroxide) in order to have a good reaction time; they have used some recirculation on the system and studied the main product by HPLC. The system has allowed a much better separation and has helped the reaction to be shifted towards the product. They have also studied different possibilities in order to reduce the amount of raw material. For this matter, the recycling of the polar phase, this allows a much smaller methanol input to the system but a good final quality biodiesel [14]. Similar works have been done by Chemseddine and Audinos [15], Othman et al. [16] and Sdrula [17]

In their work, Chemseddine and Audinos [15] used a cation exchange membrane to study the esterification reaction of oleic acid with methanol. They have found a final conversion of 80% after 8 h of reaction at room temperature. Even more, due to the way the system is presented, the final process is cleaner than conventional stirred reactors. On the other hand, the authors have also contributed with a kinetics expression based on the concentrations of reactants as well as on the protons in the fixed layer. This expression satisfactory correlates the experimental data.

When analyzing the transesterification kinetics model, as done by Cao et al. [18], it is important to notice that there are two phases involved, a moving phase, where biodiesel, glycerol and methanol will be and a non moving phase where constant properties could be assumed as well as perfect mixing. This is because in this type of reactor, as described by Cao et al. [18], the triglycerides and methanol is being recycling to the main reaction step while biodiesel and glycerol are being separated. Therefore, different concentrations and different kinetics expression needs to be developed for each compound. They have found a great agreement with their experimental data as well as corroborated that the controlling step in this system is the reactions from diglycerides to monoglycerides, as also seen by Vicente et al. [19] and Komers et al. [20]. When instead of basic catalysis, acid is chosen, Othman et al. [16] have demonstrated that an acid environment to carry on the transesterification reaction, when using polymeric membranes for the separation, is not the best scenario due to damage suffer in the membrane. On the other hand, when using basic solvent systems (pH = 8.6), it was found a considerable improvement for the permeability of the main products involved the reaction. A similar study was done by Want et al. [21] where ceramic membranes were used to purified biodiesel.

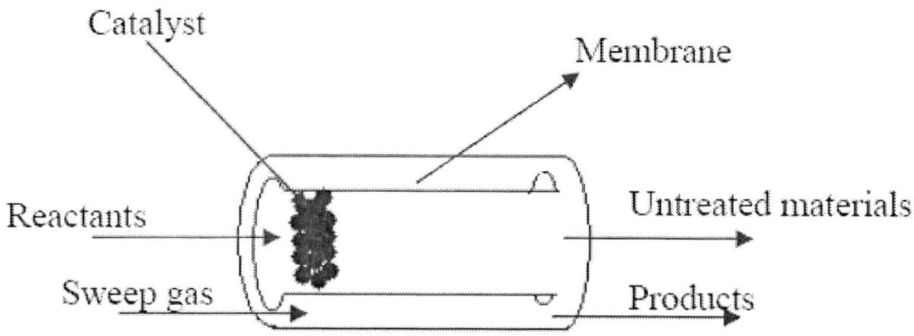

Figure 1. Schematic representation of a general membrane reactor.

As it can be seen, membranes are gaining more relevance due to its great benefits, producing a more clean process, a more environmentally friendly alternative and a very pure final product. However, besides the mathematical challenge when obtained kinetics expression, the price of this membranes is most of the time quite high. This price makes the technology less attractive from a commercial point due to the possibility of a higher final biodiesel price or a less income compared to other technologies. However, new membrane options could become available allowing the process to be competitive and the idea of a green process should be kept in mind.

Other technology that is suitable for biodiesel production is reactive distillation systems. The main idea with this technology is to have a reaction and separation step in the same equipment and almost at the same time. Several works could be found in the literature addressing this technology for biodiesel production [22-25]. Dimian et al. [22] have studied the possibility of using this technique but with two reaction sites in the column, using light and heavy alcohols to have reaction steps in the top as well as in the bottom of the distillation column. They have found that this approach is quite a compact system that is more than suitable to produce the desirable product with high yields. Similar work has been done by Kiss [23] and Kiss et al. [24]. In those work it was studied the advantages of having adsorption for separation when producing biodiesel [23], this technology not only increases the productivity but also decreases the amount of downstreaming separation required, also reducing the investment cost associated to it. Even more, when adding integrated heat analysis [24], the reduction in operational cost are also considerable reduced, improving the general concept of the system.

Finally, I would like to discuss the use of monolithic reactors for biodiesel production [10]. This technology has been used for several different areas such as for solution for environmental pollutants [26-28]. Tonetto and Marchetti

[10] have used this technology to test several catalyst, such as metals/Al_2O_3-corderite. They have tested several catalysts in the powder form in order to study the feasibility of this alternative to carry on the transesterification reaction as well as the esterification reaction, they have found promising results. From those results, the use of the best alternative as monolithic was implemented. It was found that after 6hs at 120°C the conversion into biodiesel is 59% for the transesterification reaction. The operational conditions needs to be improved in order to reach the desirable industrial conversion; however, it is important to point out that the system could present some leaching from the catalyst, allowing the reaction to take place in a homogeneous way but loosing the catalyst after the reaction and purification steps. On the other hand, powder options show a much higher final conversion, almost 90%. Nevertheless, this will mean to have a heterogeneous alternative with the advantages and disadvantages of that system, as described previously.

Although this technology is not yet in the industrial develops point, it has been also tested to carry on the esterification reaction of oleic acid in order to see how suitable this option is to treat waste raw material. Figure 2 shows that for all testes materials the conversion achieved in 6h are over 65% but not yet in the equilibrium range.

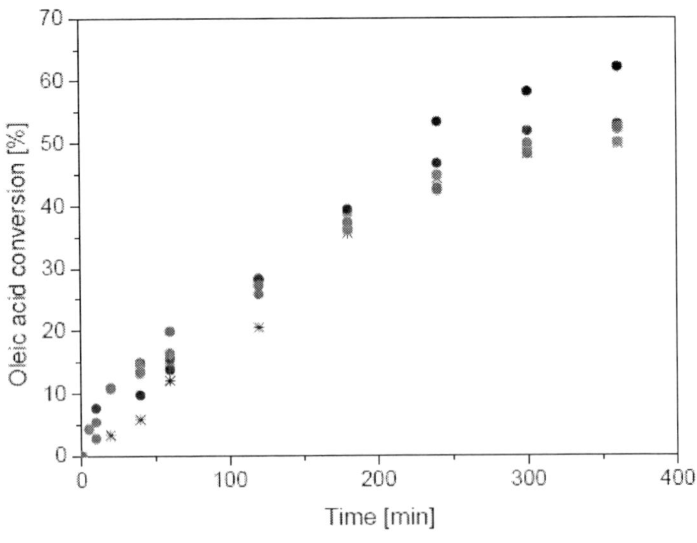

Figure 2. Conversion of Oleic acid. ● Ca/Al_2O_3, ● K/Al_2O_3, ● Na/Al_2O_3, ● Ba/Al_2O_3, ● Al_2O_3, ∗ no catalyst.

The best option is the Ca based catalyst. This results show that the reaction could be carried on in a continues process, reaching a possible higher final yield. However, this could have happened in a longer reaction time that will produce the technology to be out of the industrial perspective. Optimization on this topic is important in order to find alternatives productions technologies.

As presented in this Chapter, other technological solutions are being studied (and some others that have not been addressed in this book as well), with the aim of finding new options for biofuel productions reducing emissions, effluents and improving the process as well as the quality of the final product.

REFERENCES

[1] Ma, F. Hanna, M. A. *Biodiesel production: a review. Bioresource Technology*. 70. (1999), 1-15.

[2] Fukuda, H. Kondo, A. Noda, H. Biodiesel *Fuel Production by Transesterification of Oils. J. Bioscience and Bioengineering.* 92(5). (2001), 405-416.

[3] Barnwal, B.K. Sharma, M.P. *Prospects of Biodiesel production from vegetable oils in India. Renewable and Sustainable Energy Reviews.* 9(4). (2005), 363-378

[4] Srivastava, A. Prasad, R. *Triglycerides-based diesel fuels. Renewable and Sustainable Energy Reviews.* 4. (2000), 111-133.

[5] Marchetti, J.M., Miguel, V.U., Errazu, A.F. *Possible methods for biodiesel production. Renewable and Sustainable Energy Reviews.* 11. (2007), 1300-1311.

[6] Knothe, G., Van Gerpen, J., Krahl, J. (Eds.), *The biodiesel handbook,* AOCS Press, Champaign, Illinois, 2005, 302 pp.

[7] Schuchardt, U., Sercheli, R., Vargas, R.M. *Transesterification of vegetable oils: a review. Journal o the Brazilian Chemistry Society*. 9(1). (1998) 199–210.

[8] Dubé, M.A., Tremblay, A.Y., Liu, J. *Biodiesel production using a membrane reactor. Bioresource Technology*. 98. (2007), 639–647.

[9] Kiss, A.A. *Separative reactors for integrated production of bioethanol and biodiesel.* Computers and Chemical Engineering. (2009) Article in press.

[10] Tonetto, G.M., Marchetti, J.M. *Transesterification of soybean oil over Me/Al_2O_3 (Me = Na, Ba, Ca and K) catalysts and monolith K/Al_2O_3-corderite. Topics in Catalysis*. (2010) Article in Press.

[11] Feng, D., Wang, Y., Wang, D., Wang, J. *Pg-Ag-Au-Ni membrane reactor for methoxymethane steam reforming. Journal of Chemical Engineering Data.* 54. (2009), 2444-2451.

[12] Sá, S., Silva, H., Sousa, J.M., Mendes, A. *Hydrogen production by methanol steam reforming in a membrane reactor: palladium vs. carbon molecular sieves membranes. Journal of Membrane Science*. 339. (2009), 160-170.

[13] Yu, C.Y., Lee, D.W., Park, S.J., Lee, K.Y., Lee, K.H. *Ethanol steam reforming in a membrane reactor with Pt-impregnated Knudsen membranes.* Applied Catalysis B: Environmental. 86. (2009). 121-126.

[14] Paigang, C., Dubé, M.A., Tremblay, A.Y. *Methanol recycling in the production of biodiesel in a membrane reactor. Fuel.* 87. (2008), 825-833.

[15] Chemseddine, B., Audinos, R. *Use of ion-exchange membranes in a reactor for esterification of oleic acid and methanol at room temperature. Journal of Membrane Science*. 115. (1996), 77-84.

[16] Othman, R., Mohammad, A.W., Salimon, J. *Application of polymeric solvent resistant nanofiltration membranes for biodiesel production. Journal of Membrane Science*. 348. (2010), 287-297.

[17] Sdrula, N. *A study using classical or membrane separation in the biodiesel process. Desalination.* 250. (2010) 1070-1072.

[18] Cao, P., Tremblay, A.Y., Dubé, M.A. *Kinetics of canola oil transesterification in a membrane reactor. Industrial and Engineering Chemistry Research.* 48. (2009), 2533-2541.

[19] Vicente, G., Martinez, M., Aracil, J., Esteban, A. K*inetics of sunflower oil methanolysis. Industrial and Engineering Chemistry Research*. 44. (2005), 5447-5454.

[20] Komers, K., Skopal, F., Stloukal, R., Machek, J. *Kinetics and mechanism of the KOH-catalyzed methanolysis of rapeseed oil for biodiesel production. European Journal of Lipid Science and Technology.* 104. (2002), 728-737.

[21] Wang, Y., Wang, X., Liu, Y., Ou, S., Tan., Y., Tang, S. *Refining of biodiesel by ceramic membrane separation. Fuel Processing Technology.* 90. (2009), 422-427.

[22] Dimian, A.C., Bildea, C.S., Omota, F., Kiss, A.A. *Innovative process for fatty acid esters by dual reactive distillation. Computers and Chemical Engineering.* 33. (2009), 743-750.

[23] Kiss, A.A. *Biodiesel by reactive absorption- towards green technologies.* 19th European Symposium on Computer Aided Process Engineering –ESCAPE 19. (2009), 847-852.

[24] Kiss, A.A., Dimian, A.C., Rothenberg, G. *Biodiesel production by heat-integrated reactive distillation.* 18th European Symposium on Computer Aided Process Engineering –ESCAPE 18. (2008), 775-780.

[25] Kiss, A.A. *Novel process for biodiesel by reactive absorption. Separation and Purification Technology*. 69. (2009), 280-287.

[26] Kreutzer, M., Kapteijn, F., Moulijn, J. *Shouldn't catalysts shape up?: Structured reactors in general and gas–liquid monolith reactors in particular. Catalysis Today.* 111. (2006), 111-118.

[27] Pérez-Cadenas, A., Zieverink, M., Kapteijn F. Moulijn, J. *High performance monolithic catalysts for hydrogenation reactions. Catalysis Today.* 105. (2005), 623-628.

[28] Avila, P., Montes, M., Miró, E. *Monolithic reactors for environmental applications: A review on preparation technologies. Chemical Engineering Journal.* 109. (2005), 11-36.

Chapter 10

GENERAL CONCLUSIONS

1. CONCLUSIONS

In this Chapter 1 would like to give a short summary of several of the technologies discussed previously and compared them one more time in order to give a short summary of what has been communicate along this book.

Several options were analyzed, the main variables have been shown and how they affect the process, either improving the final yield or actually being a drawback on the general process.

A brief comparison of the major variables involved in several technological alternatives has been done and it is showed in Table 1

As it can be seen from Table 1, all the technologies, except for the super critical option could be carried out at low temperatures; this implies a low operation cost for those technologies. The presence of water and free fatty acid, two of the most common impurities, has different effects on the main yield. Free fatty acids could be esterified into biodiesel for all the technologies except for the base homogeneous catalyst. In this case soap will be produced and therefore the amount of this impurity needs to be very low. On the other hand, if enzymes do not have a proper substrate, or the resin is not strong enough, water could be a big concern. Not only the water that could be in the vegetable oil, but also the water produced by the esterification reaction, making the catalyst not suitable to treat waste oil with high amounts of free fatty acids.

Table 1. Comparison of different technologies in the biodiesel production industry [1]

Variable	Base	Enzyme	Supercritical	Monolithic	Resin	Acid
Temp. [°C]	60-70	30-50	200-350	50-180	60-180	50-80
FFA	Soaps	Esters	Esters	Esters	Esters	Esters
Ester yield	Normal	High	High	Normal	Good	Normal
Purity of glycerol	Difficult	Simple	Simple	Simple	Simple	Difficult
Reaction time	1-2 h	8 h to days	4 min	6 h or more	Variable	4 h -3 days.
Ester purification	Difficult	Simple	Simple	Simple	Simple	Difficult
Cost	Cheapest	Expensive	Expensive	Medium	Medium	Cheaper
Amount of equipment	High	Low	Low	Low	Low	High

Looking at the more economic part of Table 1, the amount of equipment required for each alternative is related to the separation and purification than to the reaction itself, reaction is carried out in either one or two reactors (few times could be in three reactors), but the downstreaming separation is attach to how easy is the separation of the catalyst from the rest. Ones the catalyst has been removed, the distillation of alcohol allow the recycling of it, and a simple decanter could be enough to separate the glycerol from the biodiesel. Therefore, as the catalyst is in the same phase as the main reactants, the amount of equipment required is bigger, meaning a more initial investment, as shown for the acid and base homogeneous process. However, those two processes have the cheapest catalyst price, being the enzymes the option with the highest catalyst price due to the preparations techniques.

Contrarily, the supercritical technology that has a quite high operational cost, due to the high temperature and pressure, has the lowers reaction time, making it quite a nice technology to carry on the biodiesel production. Another interested and new option is monolithic reactors. However, it is from our own experience that the catalyst suffer some leaching, reducing its activity, and showing that even it is a nice and improved technology, there are some technological issued that need to be solved.

When using a base or an acid in an homogeneous medium, the separation of product, either glycerol or ester, it is quite difficult, as being shown, separation, neutralization, filtration, etc are required in order to achieved the desirable purity for the biodiesel. Even more, the final glycerol does not have a pharmaceutical grade, its purity is in the industrial grade and therefore it will be sold with a lower price. And, if biodiesel production continues increasing at this rate, it could become a very cheap raw material.

Because of this is that new uses for glycerol as raw material are being discussed and implemented nowadays. Several works can be found in the literature regarding this point, as the work done by Taarning et al. [2] where the modification by oxidation in a basic medium using alcohol was established and proved to be an alternative way to modified glycerol into propanodiols that could be used for further transformations. This work showed three possible modifications of glycerol using gold catalyst in order to produce more valuable final products. All the possibilities showed high conversions and very good selectivity's towards the desirable product. A similar work was done by Voirol et al. [3] where they have studied the modification of glycerol into gas oil blend products with the aim of compared it with blends of methyl esters. They have found that Glycerol Ter Butyl Ether has no disadvantages, from a pollutant study, compared with regular diesel. Looking to the glycerol as a raw material with another perspective, to use glycerol to produce comestible oil, Bonet et al. [4] have done a work showing that glycerol could be transformed into glycerol triacetate. They have found that a reactive distillation system was very suitable to perform this modification. Much more possible technological solutions could be found for glycerol as raw material, I recommend the paper by Zheng et al. [5] for further details.

Therefore, when selecting a technology for biodiesel production, several variables are involved, from the operation of the plant, the environmental concerns, the economics involved as well as the market studies. Showing that the decision is a compromise of all them to make the best selection.

References

[1] Marchetti, J.M. *Past, Present and Future Scopes in the Biodiesel Industry. World Conference of Science, Engineering and Technology*. Oslo, 2009.

[2] Taarning, E., Madsen, A.T., Marchetti, J.M., Egeblad, K. Christensen, C.H. *Oxidation of glycerol and propanodiols in methanol over heterogeneous gold catalyst. Green Chemistry*. 10. (2008), 408-414.

[3] Voirol, A.J., Durand, I., Hillion, G., Delfort, B., Montagne, X. *Glycerin for new biodiesel formulation. Oil and Gas Science and Technology*. 63. (2008), 395-404.

[4] Bonet, J., Costa, J., Sire, R., Reneaume, J.M., Pleşu, A.E., Pleşu, V., Bozga, G. *Revalorization of glycerol: comestible oil from biodiesel synthesis. Food and Bioproducts Processing*. 87. (2009), 171-178.

[5] Zheng, Y., Chen, X., Shen, Y. *Commodity chemicals derived from glycerol, an important biorefinery feedstock. Chemical Reviews*. 108(12). (2008), 5253-5277.

INDEX

A

B

C

D

E

F

G

T

V

W

Y

Z